INDICATORS FOR URBAN AND REGIONAL PLANNING

The interplay of policy and methods

CECILIA WONG

Routledge
Taylor & Francis Group

LONDON AND NEW YORK

First published 2006 by Routledge, 2 Park Square, Milton Park, Abingdon, Oxon, OX14 4RN
Simultaneously published in the USA and Canada by Routledge, 270 Madison Avenue,
New York, NY 10016

Routledge is an imprint of the Taylor & Francis Group, an informa business

© 2006 Cecilia Wong

Typeset in Aksidenz Grotesk by Taylor and Francis Books
Printed and bound in Great Britain by TJ International Ltd, Padstow, Cornwall

British Library Cataloguing in Publication Data
A catalogue record for this book is available from the British Library

Library of Congress Cataloging in Publication Data
Wong, Cecilia.
Indicators for urban and regional planning : the interplay of policy and methods / Cecilia
Wong.
p. cm. -- (The RTPI library series)
Summary: "Examines the process of indicator development in Urban and Regional
Planning. Includes discussion of methodological debates and case study analysis of depri-
vation, urban and regional development, sustainability and planning"--Provided by
publisher.
Includes bibliographical references and index.
ISBN 0-415-27451-6 (hb : alk. paper) -- ISBN 0-415-27452-4 (pb : alk. paper)
1. City planning. 2. Regional planning. 3. Social indicators. I. Title. II. Series.

HT166.W649 2005

307.1'216--dc22

2005008574

ISBN10: 0-415-27451-6   ISBN13: 978-0-415-27451-7 (hbk)
ISBN10: 0-415-27452-4   ISBN13: 978-0-415-27452-4 (pbk)

To Mark and Viola

CONTENTS

ILLUSTRATIONS

TABLES

FIGURES

PREFACE

The ideas of writing a book on indicators was conceived when I was awarded an ESRC Research Fellowship far back in the late 1990s. With the support from Routledge and the Royal Town Planning Institute, I was given the green light to go ahead with the book. By that time, I had just moved job to Liverpool University, become a mum, and been caught in the new wave of evidence-based government policies. The last few years have been a very exciting time both in my professional and domestic life. As a result, there have been competing demands on my time and attention, which have tested my time management and task-juggling skills to the maximum. Viola is now five years old and able to sit through her parents' conference paper presentations on indicators at Grenoble and Vienna. She has made a significant contribution to the lengthy writing of this book by simply refusing to sleep through nights and exhausting her parents during the first three years. The resulting prolonged schedule of writing means that I have had the opportunity to complete in the mean time a number of research projects for different government departments and agencies, which have provided first-hand experience of the brave new world of evidence-based policy-making. During this process, I have learned a lot from practitioners and planners in the field, and have been able to be a reflective researcher to connect the technical research with the complex political and policy demands. This book provides a very useful forum for me to share these ideas.

I would like to take this opportunity to thank Andreas Bäing, Mark Baker, Bob Barr, Mike Coombes, Iain Deas, Nick Gallent, Anne Green, Philip Jeffery, Sue Kidd, David Owen, Alasdair Rae, Simon Raybould and Brian Robson, who have worked with me over numerous research projects on indicators and policy monitoring. It has been enjoyable to work with them and there is so much that I have learned from them. I would also like to thank David Adams, Chris Banister, Ian Gilfoyle, Vinant Goodstat, Peter Halfpenny, Patsy Healey, Michael Hebbert, Ian Masser, David Massey, John Punter, Peter Roberts and Gwyn Williams, and the late Michael Breheny, Valerie Khan and Moss Madden, who have given me very helpful advice and support in my research and career development. Special thanks are due to Peter Batey for bringing me to study and work at Liverpool at two different points of time, and to David Shaw for being such a wonderful head of department and offering me lifts to the train station on many dark and cold winter nights. My heart-felt thanks are also due to colleagues at the University of

Newcastle upon Tyne (CURDS), University of Manchester (Planning and Landscape) and University of Liverpool (Civic Design) for providing a stimulating and friendly working environment. I am also in debt to my lecturers at the Chinese University of Hong Kong (Sociology) for introducing me to the world of quantitative social research methods and encouraging me to pursue postgraduate studies abroad.

This book would not have been completed without the continuous support and encouragement from Robert Upton (RTPI), and the very helpful comments from Cliff Hague (series editor) and an anonymous reviewer on earlier drafts of the manuscript. Special thanks are due to Luis Riffo, it was his email from Chile that galvanised my energy to finish writing the full manuscript over Christmas 2004. I would like to thank Helen Ibbotson, Georgina Johnson, Caroline Mallinder, Jason Mitchell and Tony Nixon from the Taylor and Francis Group. They have been extremely patient and understanding of my delayed schedule and have been very effective in getting this book ready for publication. I would also like to acknowledge the support from the Economic and Social Research Council (Grant number H52427003094) for a 4-year research fellowship on the development of quantitative indicators.

Finally, but not least, I would like to thank Mark Baker, my husband and best friend, for his patience and support. As always, he has had the misfortune of being asked to read through various drafts of the manuscript as well as sharing the task of looking after Viola.

(Yin Ling) Cecilia Wong
3 March 2006

CHAPTER 1

INTRODUCTION

THE VOGUE OF INDICATORS

The need for more and more indicators has become a worldwide phenomenon since the early 1990s. A flavour of this global development can be viewed by entering the key word 'indicators' in an Internet search engine. This produces an overwhelming 5.7 million websites that are devoted to indicators, ranging from global institutions such as the United Nations and the World Bank to some local community indicators projects. If one can keep aloof as a curious observer, one may ask some cynical questions: 'To what extent can this global indicator perfusion be seen as the beginning of a trend of popularisation of statistics?' 'Are indicators just the policy accessories of the time – a *must have* in the current fashion of governance?' 'Is the current development of indicators an old hat from those developed thirty years ago?' 'Will the current hyping of indicator projects fall into an inevitable cycle of downfall in the future?'

The history of using quantitative indicators to guide policy-making can be traced back to the 1940s when the monthly Economic Indicators was first published to measure the buoyancy of the US economy. This success in developing a set of reliable economic indicators prompted US social scientists, welfare advocates and civil servants to develop indicators to measure social change in the mid-1960s. The term 'social indicators' was popularised by Raymond Bauer (1966), who was commissioned by the National Aeronautics and Space Administration (NASA) to examine the impact of the space programme on US society. The idea of compiling social indicators spread rapidly from the USA to international organisations such as the Organisation for Economic Co-operation and Development (OECD) and the Social and Economic Council of the United Nations (Horn 1993), which began to develop social accounting and reporting schemes (e.g. Stone 1971; UN Statistical Office 1975). This wave of research was named the 'social indicators movement' by Otis Dudley Duncan (1969).

The influence of the social indicators movement was also evident in the publication of social reports and compendia of social statistics in Britain, Germany and France (Taylor 1981). The publication of *Social Trends* (Cook and Martin, 2005) by the Central Statistical Office in 1970 was a reflection of a genuine British interest in constructing more sensitive measures of social conditions. Unfortunately, the initial rapid development of the movement suffered a setback in the late 1970s, due to the failure of researchers to resolve conceptual and methodological difficulties (Carley 1981). Knox (1978) identified the pitfalls in the design and construction of some indicators, including the difficulties encountered in the selection, availability and reliability of data, the problem of spatial aggregation of statistics, and problems of interpretation. Social indicators also fell into disfavour with policy-makers because they were not tailored to measure their policy concerns. More importantly, governments increasingly opted for the 'magic of the market' rather than social intelligence and became less interested in social engineering and reform (Miles 1985: 31). Thus, although the 'bandwagon effect' of research during this early period of the social indicators movement did produce revelatory and reflective work, many methodological and conceptual issues still remained unresolved.

Some 20–30 years after the initial enthusiasm for the development of social indicators in the 1960s and 1970s, another wave of worldwide indicator endeavour began in the 1990s. Innes and Booher (2000: 174) describe the latest hype in indicators research as 'the community indicators movement'. While the earlier social indicators movement was very much developed in the context of social reform and welfare at the national level, the current resurgence of interest in indicators has been largely stimulated by broad environmental concerns related to creating sustainability and quality of life indicators at all spatial scales. The call for suitable 'indicators of sustainability' to provide a solid base for decision-making at all levels was explicitly stated in Agenda 21 (UNCED 1992: Ch. 40.4) of the 1992 Earth Summit conference in Rio de Janeiro. The 1996 Habitat II conference in Istanbul further reinforced the importance of community-based indicator projects to guide and track the progress towards achieving sustainability. This new environmental agenda has not only brought with it a need to employ indicators as a key mechanism for assessing environmental impact and capacity (Maclaren 1996; Macnaghten et al. 1995), but has also spurred local action and broadened concern to encompass the wider community-based issues. Since then, the magic dust of indicators has been showering every corner of the world under different banners and logos such as 'sustainability indicators', 'quality of life indicators' and 'performance indicators' (see Sawicki 2002; Swain and Hollar 2003). Some of the well-publicised community indicators projects include Sustainable Seattle (pioneered in 1993) and the Jacksonville Community Indicator Projects in the USA.

The latest comeback of indicators may suggest that there are some intrinsic values of indicators and that they will come and go with the policy current. This calls for research to underpin the reason for the pendulum of indicators to swing in and out of the policy circle in such a big way, and to identify the underlying strengths and innate tensions that orchestrate such dramatic shifts over a period of two to three decades. There is also a need to adopt a more grounded approach to ascertain the reality of the development and practice of indicators as a policy instrument.

THE EPISTEMOLOGICAL POSITION OF INDICATORS

Indicator as a word is fairly comprehensible to most people. To indicate is to 'point out, make known, show; state briefly, be a sign of, betoken; suggest, call for' (*The Concise Reference Encyclopedia and Dictionary* 1987: 850). To put it simple, indicators are statistics that provide some sorts of measurement to a particular phenomenon of concern. As with any large-scale quantitative research, indicators tend to be seen as part of the empiricist or positivist tradition. This perception, rightly or wrongly, has elevated indicators to a darling position in the policy world, as their hard and fast nature serves well as an instrument to justify and rationalise resource distribution. In order to explore the epistemological position of indicators, the analysis has to be focused on both the theoretical level and the practical reality of what indicators are and how they are developed.

The development of indicators in the 1960s had been a multi-disciplinary activity that engaged both academics and practitioners. It is, therefore, not surprising to find that different emphasis has been placed on the definition of indicators (Carley 1981; Horn 1993; Innes 1990). Three definitions are given in Table 1.1, each of which represents a particular viewpoint over the nature and purpose of indicators. Carlisle's definition highlights the fact that indicators are the result of operationalising abstract concepts of social and policy problems (Carley 1981), and they offer a guide indicating how a particular issue is structured or is changing (Miles 1985). This theoretical view serves the empirical school of thought well, as the emphasis is on finding a way to provide measurement of concepts and problems. The definition of Bauer (1966), however, adds a normative dimension to the empirical formula. This normative emphasis suggests that indicators would be used as a yardstick to measure progress and goal achievement. The emphasis on value and goal setting – the presupposition of certain innate knowledge – to benchmark against the measured result shifts the epistemological basis closer to that of the rationalists. The theoretical expression of these two definitions shows that indicators do not sit comfortably on either the empiricist or the rationalist tradition in their pure form.

Under the definition of the US Department of Health, Education and Welfare (1969), indicators are subject to another epistemological turn as it highlights the importance of interpretation. This means that value judgement would be involved in viewing some effects as better or worse. More importantly, the norm of assessment is susceptible to change and interpretation. Following neo-Kantianism, this definition opens the argument of relativism, and stresses the importance of inter-subjective communication and interpretation of meaning. This also points to the underlying tension of indicators as policy instruments, which are subject to the politicisation of interpretation and the possibility of manipulation even at the measurement stage through the choice of indicators, data sources and methods. The value-laden aspect of indicators clashes with both the rationalist and the empiricist ideology, as the foundation of securing objective knowledge from belief, opinion and even prejudice is somewhat less convincing.

Taking these early ideas on board, it is clear that the value of indicators as a form of knowledge is grounded on its methodological process of moving from abstract concepts to more specific and concrete measures to yield policy intelligence. Nonetheless, it is this unique blend of technical and normative rationality that makes indicators a 'Jekyll and Hyde' character.

Besides the debate over the nature and purpose of indicators, there are also divided views over the scope and coverage of indicators. Some authors deliberately demarcate social indicators from their economic cousins to repudiate the view that social indicators only play an auxiliary role to their economic counterpart (Cazes 1972). Such a distinction is not very helpful because it fails to clarify the sphere of social indicators. Bauer (1966: xvi) thus defined the term 'social' as 'societal' and saw social indicators as indicators for society. This broader notion of social indicators aims to take into account both social and economic considerations and to integrate them into policy decisions. This was further advanced as a line of argument by MacRae (1985: 9), who put the view that 'policy indicators' should be all inclusive, and that 'economic indicators should be joined in a single inclusive domain with the social and related to them. Indicators based on the natural sciences' he considered (ibid.), 'should also be treated in this common framework'. It is this wider notion of indicators that the discussion in this book follows.

Table 1.1 Some definitions of indicators

'Social indicators are the operational definition or part of the operational definition of any one of the concepts central to the generation of an information system descriptive of the social system.' (Carlisle 1972: 25)
'Social indicators – statistics, statistical series, and all other forms of evidence that enable us to assess where we stand and are going with respect to our values and goals, and to evaluate specific programs and determine their impact.' (Bauer 1966: 1)
'[S]ocial indicators are statistic(s) of direct normative interest & a direct measure of welfare and is subject to interpretation.' (US Department of Health, Education and Welfare 1969: 97)
'[A] set of rules for gathering and organising data so they can be assigned meaning.' (Innes 1990: 5)

THE POLITICAL-MANAGERIAL PERSPECTIVE OF INDICATOR DEVELOPMENT

The tide of indicator development in the 1960s was very much grounded on the grand ideology of social reform and welfare at the national level to remove the gap of development between the developed and less developed world. Following the collapse of communism in the 1980s, the mood of top-down social reform has been less enthusiastic. Instead, it has been replaced by another form of ideology – the emphasis on global environmental concerns. The slogan of sustainable development, advocating the importance of inter- and intra-generation equality and equity in terms of environmental resource consumption, is shouted loudly at different corners of the world. It is the bottom-up nature of the sustainability agenda that triggers the 'community indicators movement'. Whilst the approaches to the 1960s and the 1990s indicators movement are somewhat different, they both share the value of improving the quality of living of people and places. In spite of the blossoming of grassroots indicators, international organisations such as the World Bank (1995, 2003), the Asia Development Bank (Westfall and de Villa 2001) and the United Nations (UNCSD, 1996) are still actively pursuing national and urban indicator sets to measure different aspects of development.

While the grandiosity of social indicators and sustainability indicators represents significant symbolic forces in pursuing higher-order goals and values, there are other less exciting indicator sets developed by bureaucrats and government statisticians to provide day-to-day management and monitoring of policy regimes. With the ever-deepening crisis of state finances and the diminishing capacity of central government to deal with the gravity and the magnitude of unemployment and industrial decline (Stöhr 1987; Altvater 1992; Blakely 1994) in many developed nations, there has been a shift in the policy regime in the UK and the USA since the 1980s. The rise of neo-liberal ideology, commonly identified with the Reagan and Thatcher experiments, seeks to replace formalised legal regulation with market mechanisms (Dunford 1990; Sayer 1992) by emphasising supply-side, free market and competitive relations between firms and between places (Blakely 1994; Peck and Tickell 1995). This new wave of policies signifies the demise of the traditional 'Keynesian' policy models and the Fordist mode of mass production and capital accumulation (Altvater 1992). The tightening of public expenditure also means that central government has been closely scrutinising public programmes and monitoring the effectiveness of individual policy activities. Indicators are thus employed as a managerial toolkit for area targeting, performance and procedural monitoring.

In contrast to the diminishing capacity of national governments to tackle the problems of structural change effectively, international/supranational institutions such as IMF, the G7, the World Bank and the European Commission have become increasingly powerful (Held 1991). This tendency of the displacement of many of the national state's former powers and functions to international bodies, and the increasing number of initiatives and activities at the local scale, have increasingly subsumed the influence of national governments. This 'hollowing-out of the nation-state' phenomenon described by Jessop (1993, 1994) represents a new spatial order in economic geography as 'the global–local nexus'. The one result of the formation of a Single European Market is to remove the longstanding factor of nationalism to bring about economic and social integration of its constituent countries. The institutional structures and funding initiatives of the European Union, for example the European Structural Fund and Social Fund, are operated on a regional basis and stress the importance of partnership as an implementation mechanism. Spatial targeting and co-ordination of European aid has become more important since the move towards Integrated Development Operation Programmes in the 1990s (CEC 1991). The pertinence of indicators in programme monitoring and evaluation was made explicit by the European Commission (2000a) when launching its New Programming period in 2000. A guidance paper was prepared by the Commission to provide a frame of reference on how to harmonise the usage of indicators in the evaluation of a whole suite of EU structural assistance and to develop more pragmatic and flexible monitoring systems. The changing face of cities and regions, and the development of new forms of institution and governance at different spatial levels have set in train a very dynamic policy agenda. The political-managerial needs of having some forms of quantifiable measures to justify resource allocation have no doubt boosted the importance of statistics and indicators in the policy arena.

Against the canvas of the global–local nexus of institutional change, it is logical to expect a trend of decentralisation of indicator collection and usage similar to that observed in the USA (Innes and Booher 2000). The evidence witnessed in Britain is, however, quite the opposite. Although there has been a continuous process of delegation of power to local and regional actors to carry out monitoring and data collection, it is ironically undermined by the ever-strengthening process of centralisation of funding and performance control under the guidance of central government (Wong 2000). This concomitant centralisation and decentralisation tendency of policy monitoring makes Britain an interesting case study for more in-depth examination of the ways indicators are developed and used in the policy process.

Following the demise of the social indicators movement, the interest in developing social indicators was also on the wane in Britain. Indicators research had been largely neglected from the late 1970s through most of the 1980s, due in part to the wider shift of the academic paradigm from the so-called quantitative revolution to a political-economy approach (see Hall 1983). Indicators, nevertheless, continue to be used in the British policy circle as a funding allocation mechanism by identifying priority areas for all sorts of policy programmes. Central government has generally adopted a top-down approach to resource allocation, mainly via consistent and systematic analyses of a variety of chosen quantitative indices to measure the relative level of local needs. One well-known example was the Department of the Environment's 1981 Deprivation Index (DoE 1983), which consisted of eight indicators from the 1981 Population Census. The rankings of these eight indicators were then used to identify Urban Programme Areas for the eligibility of various physical regeneration grants. The Housing Corporation (1988) also derived a 'Housing Needs Index' from Population Census data to reflect relative housing needs and to allocate regional capital expenditure accordingly. There has also been a long tradition of using unemployment rates, migration levels and industrial growth statistics (DTI *et al.* 1993, 1999) to designate 'Assisted Areas' where regional aid may be granted to promote investment and business innovation.

The needs base assessment approach has, however, given way to a more entrepreneurial style of policy-making towards the end of the 1980s. The assertion of the critical role of market forces in shaping policy bears implication on the role played by indicators in the policy process, especially in relation to urban regeneration and local development. This in part reflects the rise of the so-called 'new localism', involving local actors who are increasingly aware of their role in shaping the development of the local economy (Roberts *et al.* 1990). In the face of intensified international competition, individuals, businesses, local authorities and other actors have shifted their reliance on national government policies to their own local initiatives to harness the distinctive advantages and opportunities possessed by each area to enhance effective competition (Fielding and Halford 1990; WMEB 1993). At the same time, with the tightening of public expenditure, central government started to impose more rigorous monitoring requirements of policy activities and to improve policy targeting and co-ordination (Audit Commission 1989). The monitoring culture was exacerbated by the need of spatial targeting and co-ordination of the EU funding regime. It is against this background that different government departments and public agencies have commissioned or carried out indicator studies to assess the potential and problems of different areas, in order to improve their understanding of the policy operating environment and to facilitate policy targeting (e.g. Coombes *et al.* 1992, 1993a, 1993b, 1995; Dunn *et al.* 1998; Noble *et al.* 1999, 2000a; Robson *et al.* 1995, 1998).

The funding allocation mechanisms to local development in Britain went through some dramatic changes towards the end of 1980s. Since the introduction of Estate Action in 1986 and City Challenge in 1991, resources for urban regeneration and local development have been allocated through a competitive bidding process. This fund raising game has subsequently been introduced to a suite of programmes including the Single Regeneration Budget, Rural Challenge, Regional Challenge, Local Challenge and Capital Challenge. The competitive ethos of 'challenge' funds contrasted markedly with the thrust of earlier approaches, many of which aimed to apply a consistent and systematic approach, mainly via a variety of quantitative indices, to gauge local needs. Under the competitive bidding arrangements, central government was less inclined to take a global overview of social and economic needs of different localities. Instead, there was a shift to the neo-liberal philosophy that resources were allocated to local areas through the operation of quasi-market forces. The open competition of development funds had provided extra impetus to local coalitions to make use of statistical information to advocate their case and target programme areas in their bidding documents.

In the early 1990s, a new consensus emerged within the planning circle to pay more attention to conservation and environmental protection. This has been strongly driven by the 1990 White Paper *This Common Inheritance* (HM Government 1990) and the subsequent *Sustainable Development: the UK Strategy* (DoE 1994) that was, itself, a response to Agenda 21 adopted at the Rio Earth Summit. This new environmental agenda has brought with it a need to employ suitable indicators to track progress towards sustainability. Subsequent enthusiastic responses can be found in nearly every published report (e.g. Countryside Commission *et al.* 1993; DoE 1993; Arup Economics and Planning 1995). This blossoming of sustainability indicators, however, has not been subject to as much government co-ordination across the national, regional and local spectrum as might have been expected in the early years (Stewart 1995a). Unlike the examples shown in the USA, the scale of community indicators projects is much smaller and is largely led by local authorities with input from local communities. The most well-documented examples include the pilot projects in six local authority areas carried out by the Local Government Management Board (LGMB 1995) and the approaches adopted by Lancashire County Council (Macnaghten *et al.* 1995). However, as recent research shows, the initial enthusiasm for Local Agenda 21 has not endured (Blowers and Young 2000). Most of the sustainability indicator sets in Britain were initiated by national government or agencies that have an environmental focus or remit. In recent years, there seems to be a consensus of adopting the official indicator sets (see Wong *et al.* 2000) developed by central government at both regional (DETR 1998a) and local (DETR 2000a) levels.

The most recent push to the use of indicators in Britain is closely related to the evidence-based policy ethos of the government and the belief in consumer choice and transparency. The Labour government has lifted the trends of using indicators to justify funding applications and to carry out policy monitoring and evaluation to new heights. Since its election in May 1997, different sets of indicators have been introduced in its urban regeneration (e.g. DETR 1998b, 2000b), regional development (e.g. DETR 1999a; DTI 1998) and regional planning (e.g. DETR 1998c; ODPM, 2002a, 2005a), environmental and land use planning (e.g. DETR 1998a, 1998d, 1999b, 2000c; ODPM 2005b) policies as well as the general public service delivery (e.g. DETR 1999c). A strong belief has emerged from the government that good-quality information is 'a corner-stone of democracy and . . . [is] essential to good public management and accountability' and 'offers an authoritative and impartial picture of society and a window on the work and performance of government itself' (ONS 1998: para. 1.1).

The nature of the emerging new wave of indicators research in the British policy community is different from that of the earlier social indicators movement. Rather than aiming for comprehensive and precise statements on the welfare state of the society, the emphasis has been much more pragmatic and policy-driven, and far from the intention of creating a grand-model indicator system. This pragmatic approach tends to focus on specific policy concerns and aims to provide intelligence and analysis to meet with changing policy agendas and needs. Thus, the indicators measured are not necessarily output-oriented (which is desirable for social accounting), but include input-oriented and outcome measures to enhance the understanding of the development process and the derivation of regeneration strategies.

THE INTERPLAY OF POLICY AND METHODS

The underlying ethos of governance and policy-making has been shifted and changed over the last few decades to cope with the dynamic process of socio-economic change brought by globalisation and other forces. The use of indicators as a policy instrument has thus been cast, shaped and applied at different contexts. The drive of indicator development largely comes from the emerging policy agenda and is shaped by the institutional-managerial culture of the time. The actual delivery of any robust and valid indicators is, however, highly constrained by the techno-methodological dimension of development. The plethora of policy documents published in the last few years has fuelled public discourse with concepts such as 'economic competitiveness', 'sustainability', 'urban renaissance' and 'quality of life'. The measurement of these concepts is underpinned as well as being undermined by the technical capacity of methods and data. There is a tendency for politicians to make commitment over the monitoring of policy before they work out the exact meaning of the policy concept and the technical capacity to support such a commitment. When policy demand exceeds the tech-nical capacity of development, it is inevitable that tension emerges – which in turn undermines the methodological rigour and the policy relevance of indicators.

It is this tension that imposes political pressure on government departments to invest resources on collecting new data and releasing more relevant data sources to improve the technical dimension of indicator research. This is clearly stated in the Allsopp review of statistics for economic policy-making that,

> devolution of economic policy also requires the devolution of budgetary resources – which, if the resources involved were to increase, would be likely to pose an increasing challenge to the statistical services to make sure that the data to underpin that process are fully credible.
>
> (Allsopp 2004: 1)

The move towards an information-demanding approach of policy-making and the emphasis on finding out the longer-term policy impacts point to an overhaul of the research and technical capacity of the policy machinery. More importantly, space and policy tend to be closely bound up when developing indicators are applied to the urban and regional planning field. The emphasis on the spatial dimension of indicator development has posed a series of methodological issues and dilemmas when developing indicator frameworks.

It is the interplay of instrumental rationality and normative policy context in the process of indicator development that forms the central tenet of this book. It is fair to say that most commentators who wrote about the political use of information are not directly involved in the technical and methodological aspects of indicator development to provide an all-rounded view on the nature and approaches of indicators research. The discussion made in this book is grounded on the observations made by the author who has been practising in the field of indicator research with various government departments and agencies for over a decade. The analysis aims to offer a particular set of lens to explore how the political and institutional setting of planning-related policies shape the scope, methods and interests in indicator research, and vice versa. In order to provide some reflective ideas, Britain (especially England) is used as a laboratory to reflect on the author's own research and practice experience, though indicator development elsewhere will also be drawn upon in the discussion.

The changing political agenda in Britain has led to four major trends of development. First of all, the use of indicators as a resource allocation tool in a number of policy areas has extended to the full spectrum of government activities. The second trend observed is the concomitant centralisation of monitoring guidance and decentralisation of data collection responsibility. The third area of development is a gradual shift of policy monitoring from the emphasis of outputs and implementation to the wider impact and strategic outcomes to develop an evidence base for policy-making. Finally, the construction of indicators has moved away from the simple, transparent approach to more complicated, statistical modelling methods, and the number of indicators used has been largely increased. This book aims to provide an in-depth exploration of these trends by structuring the discussion around three key themes: usage, methods and case studies. The overall conceptual framework is built upon many longstanding issues and ideas developed during the social indicators movement. In spite of the fact that planning indicators tend to be spatially oriented, many concerns raised in the 1970s merit further exploration against the contemporary policy context.

STRUCTURE OF THE BOOK

This book is structured into three parts. The first part of the book aims to explore the under-researched issues surrounding the *usage* of indicators in policy-making. Issues over the utilisation of indicators are explored through the roles played by the national statistical service, central government and local policy-makers over data production and information usage. The second part of the book focuses on the conceptual, methodological and analytical issues concerning the *measurement* of indicators. With the resurgence of academic and policy interest in using indicators to inform planning, this book reviews the latest research and approaches in respect of the measurement of indicator sets, and assesses the extent to which progress has been made. The final theme of the book is to use three broad groups of indicators to highlight the issues raised in the earlier sections on usage and measurement. These three broad groups of *case study* indicators cover the measurement of deprivation; urban and regional development; and sustainability and planning policy performance. They are chosen not only because of the knowledge and experience the author has with these indicators, but also due to their importance to policy-makers in contemporary planning. These indicators help to provide concrete examples to illustrate the key issues that have emerged from the interplay between policy and methods in current practice of developing and using indicators.

Following this introductory chapter, Part I consists of three chapters on the usage of indicators in the policy-making process. Chapter 2 provides a discussion over the relationship between theory, measurement and policy-making, and the roles played by social scientists in the development and applications of indicators. It also outlines the key institutional and managerial issues involved in the use of indicators in public decision-making. Chapter 3 explores the changing political ideology and government ethos over public expenditure and policy monitoring in the last 20 years and examines how these policy changes have led to some dramatic shifts over the pragmatic approach that has evolved around the application of statistical indicators. It then turns to discuss the response, both in terms of attitude and capacity, of local policy-makers towards the use of indicators to detect whether there has been a culture shift of using indicators in policy-making. Chapter 4 then discusses the organisation and management of statistics and assesses the adequacy and openness of the national statistical infrastructure in supporting an information-demanding policy regime. It also explores the emerging models of providing regional intelligence for decision-making.

Part II of the book has three chapters discussing the conceptual, methodological and analytical issues of indicator development, and how the technical side of indicator research shapes and is being shaped by the policy agenda. Chapter 5 highlights the critical issue of data availability and quality by identifying the inherent problems in current public data compilation practice and the latest trends of development. Chapter 6 explores alternative approaches used to improve the interpretability, analysis and presentation of indicators. It also revisits the longstanding debate over the techniques used to simplify indicator values and the pros and cons of creating composite indices. Chapter 7 explores the integration of the three key components of indicator construction (that is, policy context, theoretical perspectives and methodological issues) and revisits previous debate on the methodological process of indicator development.

Part III of the book is a case study-based discussion through three broad groups of indicators. Chapter 8 examines issues in relation to the methodology and utilisation of deprivation indicators. This is conducted by examining a whole range of indicators developed by Central Government and others since the Department of the Environment's 1981 Deprivation Index. Chapter 9 focuses on issues in relation to the methodology and utilisation of indicators that gauge urban and regional development potential. It also examines the progress in developing the intelligence and institutional structure to support the devolution of regional economic policy in England. Chapter 10 examines the development of sustainability and planning policy indicators in Britain and within the wider context of development made by the United Nations, European Union and other countries. The discussion focuses on the dilemma and problems encountered in developing indicators with a consensual and all-inclusive approach.

The final chapter of the book provides a synthesis of the key issues in relation to the measurement and the usage of indicators in urban and regional planning, and identifies a number of pointers to set the agenda for future research and development.

PART I

INDICATOR USAGE AND

POLICY-MAKING

CHAPTER 2

INDICATORS AND POLICY-MAKING

Following on from the epistemological discussion of the nature of indicators in Chapter 1, this chapter further examines the relationship between theory, measurement and public policy-making. It then identifies the dilemma and tension faced by social scientists, and the role they can play in indicator research. It concludes by outlining some longstanding institutional and managerial issues involved in the use of indicators in policy-making.

THEORY, MEASUREMENT AND POLICY-MAKING: THE TANGLED TRIANGLE

The debate over the nature and purpose of indicator research has largely been focused on two dichotomies: theoretical versus empirical, and basic scientific versus valuative. The relationship between theory and measurement has long been a subject of debate. In the mid-nineteenth century, Auguste Comte (1844: 25) condemned observation without theory as 'empiricism' and theory without observation as 'mysticism'. The contention between empirical measurement and theoretical ideas is strongly manifested in indicator research. The empiricist holds the view that data collection comes first and working out its meaning comes later, while the theorist insists on having some sort of *a priori* theoretical model to guide the selection and interpretation of data. As a matter of good practice, the advice from most social research text is that measurement should be guided by theories (e.g. Babbie 1992; Bulmer 1977) to avoid amassing data without giving precise definitions to guide policy action (De Neufville 1975; Fox 1974). After examining a series of James Coleman's reports on education, MacRae (1985) criticised them as empirical studies set off from common sense, which could easily lead to bias and be manipulated by decision-makers.

While the theory–data nexus is widely accepted as the norm, there is also a realisation that the view of a one-way, linear relationship between theory and data has been oversimplified. Ragin (1987) argued that a divide between theories and concepts, on the one hand, and data gathering and analysis, on the other, tended to undermine the potential of the data collected. For pragmatic purposes, it is inevitable to find that theory and measurement mesh in an iterative loop during the process of indicator development. As discussed later in Chapter 5, one of the major millstones of indicator research is the lack of robust and reliable data. Hence, Ragin's assertion that initial examination of data usually exposes the inadequacy of theoretical formulations, and further data analysis can lead to progressively more refined concepts, offers a more realistic description of the actual practice of indicator research.

Turning to the other dimension of the argument, the focus is on whether indicators should be policy-related or scientific measurements of social change (MacRae 1985). As discussed in Chapter 1, the value-laden policy dimension of indicators clashes with both the rationalist and the empiricist ideologies. The mix of objective measures and normative policy action makes indicators an enigma in social research. In order to defend the scientific purity of indicators, many social scientists make it explicit that indicators should be used to advance the state of social theory and have to be explored from a theoretical basis and in the context of causal social models (Sheldon and Moore 1968; Land and Felson 1976; Land and McMillen 1980). However, if statistics and indicators are used to serve public debate and policy action, then they have to be more than pure research tools of analysis (MacRae 1985; Miles 1985).

The discussion so far suggests that the dichotomy of theory and empirical measurement should not be overstated and they should be treated as two sides of the same coin in indicator research. Opinions regarding the purpose of indicators as tools for policy or scientific domains are somewhat diverse. The demarcation between the two sets of concern again may be more of an intrinsic academic debate rather than what matters in reality. The phenomenon and concept to be measured in many cases is not a static object but a moving beast. Scott Greer's (1969) discussion of the changing nature of problem definitions over a period of time best explains such a dynamic process in policy circles. He argued that a public problem couching on folk frame of reference will soon develop to a policy problem through seeking solutions, and after systematic inquiry it will in turn become a scientific problem. If policy context and experience is, as proposed by Innes (1990), treated as a form of knowledge, then the boundary between the two will be blurred. In recent years, many indicator projects set out to measure concepts such as 'sustainable development', 'economic competitiveness' and 'public service delivery', which are the outcome of evolving policy discourses. One recent example is the publication of the *Sustainable Communities* document by the Office of the Deputy Prime Minister (ODPM 2003a) to tackle the deepening housing crisis across different parts of England. While the badge of 'sustainable community' is sweeping the policy community to advocate neighbourhood and housing renewal, there is still little understanding of what it exactly entails. The Urban Policy Network of the ODPM, therefore, commissioned a detailed analytical report (Kearns and Turok 2003) to unpack the key elements and characteristics of a sustainable community in order to brief its civil servants and other policy-makers. The findings of the report subsequently found their way into the Egan Review (Egan 2004) of skills required to push the agenda of sustainable communities. It is also clear that policy discourses such as 'polycentricity', 'spatial planning' and 'social exclusion' developed from the European Union tend to find their way very quickly into British policy documents, with research and clarification coming later. These examples lend strong support to Greer's formulation and that policy context and experience should be accorded some weight on a par with theoretical knowledge to guide the design and measurement of indicators.

Notwithstanding the reality that theory, measurement and policy is closely intertwined with one another and often found as a tangled web of relationships in the process of indicator development, the axes of theory–data and basic research–policy application do offer a way to examine the nature and emphasis of different types of indicator research. Based on these two dimensions of debate, a four-fold classification of indicator research could be developed (see Table 2.1). Type I and Type II of the classification are linked to the apolitical path of basic research and fit well with the pure epistemological positions. The remaining two groups have a strong policy focus, with Type III grounded in the importance of theoretical frameworks and Type IV driven by empirical data. It is fair to say that many indicator sets developed for policy monitoring are closely following the Type IV protocol. One well-known example is the Department of the Environment's (DoE 1983) 1981 Z-score deprivation index. Other examples of this approach include consultancy research on local economic development and economic competitiveness (e.g. WMEB 1993; Pieda 1995). However, it is the valuative-theoretical approach (Type III) that has been widely advocated by researchers (e.g. Coombes and Wong 1994; De Neufville 1975; MacRae 1985) to guide indicator development.

INDICATORS AND POLICY PRACTICE: UNDERSTANDING, INTERESTS AND VALUES

The understanding of both theoretical ideas and policy contexts is of prime importance in the process of indicator development if indicators are used to inform policy decisions. The discussion here is based on an Economic and

Table 2.1 A four-fold classification of approaches to indicator measurement

	Theoretical	Empirical
Basic, scientific	I	II
Valuative, policy-oriented	III	IV

Social Research Council (ESRC)-funded study conducted by the author between 1995 and 1998. This research aimed to provide an in-depth study to tackle the problems encountered in developing indicators to inform local economic development (LED) decisions. The overall research design impinged on the integration of three key research components: policy, theories and methods. The project started off with a major literature review to derive an extensive list of key factors that are considered to be important to LED (Wong 1998a). Primary data was collected both through postal questionnaires and in-depth interviews, so as to examine practitioners' views on LED and indicator development and usage in two case study areas, the North West and the Eastern Region (see Wong 1998a, 1998b, 2000). These two regions were chosen because of their contrasting socio-economic conditions and experiences that would provide the conditions for a more robust interpretation of the findings. An extensive data collection exercise was carried out to compile a full dataset containing sixty-one LED indicators as well as the associated explanatory documentation. However, after an initial exploration and sensitivity testing of their statistical properties, only twenty-nine of these indicators were included in the final analysis. Multivariate analyses were then carried out to examine the structure of relationships among the compiled LED indicators (Wong 2001, 2002a) and the relative strengths of relationship between the LED indicators and various performance variables (Wong 2002a).

Despite the fact that there was not a specific policy client for the ESRC project, a significant data collection exercise was carried out to elicit views from policy-makers on LED issues, both through questionnaire surveys and in-depth interviews. The reason for doing so was two-fold: first, to corroborate and validate the theoretical findings from literature with empirical evidence from practitioners, and, second, so as to adopt the value relevance approach of Max Weber (1964), namely, to penetrate the subjective meanings that actors attach to their own behaviour and to the behaviour of others to improve the explanatory understanding of issues surrounding LED. After a review of literature, eleven factors considered to be important to LED were identified. The survey response of practitioners (see Table 2.2) confirmed these as comprising an exhaustive list. While there is a high level of stability and consistency over the perceived importance of different factors, this conclusion also highlights the fact that their relative importance is circumstantial. Even when a factor emerged as common across the two regions, the reasons behind the assigned importance can be quite different.

The empirical evidence subsequently collected from in-depth interviews debunked the logistics and myth behind the fantasy of high-tech development and the quality of life syndrome (Wong 1998a). For instance, the type of employment provided by high-tech development did not necessarily match the skills of the local residents, who tended to be semi- or unskilled labour in inner-city areas; hence, this factor was not highly regarded in the North West. The attitude towards research and development in the Eastern Region was positive but also cautious. The high-tech image projected from the Cambridge area did not seem to stimulate massive enthusiasm as, for instance, the success of the Cambridge Science Park was not seen as a realistic role model for others to follow. Equally interesting were the different views over quality of life and LED. Instead of being a contributor to the return of business investment, quality of life was seen as the consequence of prosperity. Furthermore, it was widely agreed that there were always some decent pockets of residential areas within commutable distance wherever one worked in Britain. The obvious inter-regional dimension over the ranking of quality of life in the questionnaire data was subtly shown in the interviews, in that those in the Eastern Region were more aware of the good living quality they had. Their counterparts in the North West were, nevertheless, conscious of the fact that there were many affluent suburbs and scenic Cheshire villages from where residents could commute to work elsewhere in the region.

The findings from literature and policy-makers were further tested by the assembled indicator database. Statistical findings from principal component

Table 2.2 Mean rank of local economic development factors

North West (n = 73)			Eastern Region (n = 64)
Traditional factors			
Physical factors	2.82(1)●	●2.97(1)	Location
Location	3.37(1)●	●3.39(0)	Physical factors
Human resources	3.50(0)●	●3.88(2)	Infrastructure
Finance and capital	4.07(4)●	●4.16(1)	Human resources
Infrastructure	4.85(1)●	●4.67(5)	Finance and capital
Knowledge and technology	6.07(4)●	●5.57(8)	Knowledge and technology
Industrial structure	6.75(8)●	●6.08(14)	Industrial structure
Intangible factors			
Institutional capacity	5.56(4) ●	●5.63(8)	Quality of life
Business culture	5.68(12)●	●6.09(9)	Institution capacity
Community identity	6.77(11)●	●6.22(11)	Business culture
Quality of life	6.84(5) ●	●6.72(12)	Community identity

Source: Wong 1998a: 711

Note: The mean rank values of LED factors were calculated according to their importance given by the respondents in the survey; low value of mean rank implies greater importance of the factor. The value in the brackets is the number of respondents who believed that a factor was not at all important.

analysis (Wong 2002a) echo the views expressed by the majority of practitioners (Wong 1998a) that traditional factors such as location, infrastructure, finance and human resources are more important in the process of LED. The statistical analysis also lends more support to the views of academics such as Doeringer *et al.* (1987) on the importance of having a favourable industrial structure than to the views of practitioners in the two English regions. It is also interesting to note that the human resource factor tends to be frequently associated with the intangible factors of quality of life and business culture. This result again mirrors the findings of the in-depth interviews with practitioners (Wong 1998a) that skills and qualifications of human resources are widely perceived as important factors in successful LED.

The experience gained from this study shows that the engagement of policy-makers in the process of indicators development can serve two main purposes. The first is to enhance an understanding of the policy operation environment, the subjective values and interests that policy-makers have over the research; the other purpose is that of providing for the ultimate decision over policy choices. With regard to the former, a better understanding of the policy context and values will enhance the focus of research and provide clarity to the concept to be measured. More importantly, it will provide a frame of reference when policy recommendations are made after reporting the research findings. This viewpoint is shared by Martin that in order to influence the influentials, it is important to incorporate policy-makers' views and needs by 'determin[ing] the considerations and standards of evidence that they [policy-makers] will accept or require. For the most part, these concerns will not conflict with the research requirements for validity and reliability' (Martin 1989: 48).

With regard to the issue of ultimate decision-making, it is clear that policy choices and decisions are value judgements, which cannot be scientifically determined by statistics. More importantly, it will be naïve to assume that statistics and research analysis bear any direct relationship to policy decisions. The account given by Shirley Williams, a politician with a close association to academia, on the relationship between policy and research sheds some light on this debate:

> Obviously a policy maker who is elected will have his or her own strong views. Those views will be shaped by the commitments already entered into the manifestos of his or her party; they will be influenced to a great extent by personal principles and prejudices; and the views will be modified by the policy maker's awareness of pressure from colleagues and what colleagues are likely to accept or reject. If the policy maker is a minister in a department, then his relations with other ministers and other departments within the government structure will influence his decisions. So will the estimates he makes of the interest groups whose

help he needs to carry new policies out: will they cooperate or will they resist? Then Parliamentary opinion and public opinion will have to be taken into account. Finally, the policy maker will have to assess the need for policies, their possible effects and their costs. All this relates to any policy decision: yet of that long list only two, namely the assessment of the need for, and probable effects of, the policy and of its cost, are clearly related to research.

(Williams 1980: 2)

A fine line has thus to be drawn between role of the research analyst and those who are responsible for making decisions should they be politicians, government officials or local communities. This issue inevitably links to the debate about the role of researchers in the process of indicator development and the concern over the infiltration of policy values and interests into research findings.

THE ROLE OF SOCIAL SCIENTISTS IN INDICATOR RESEARCH

Indicators as a set of statistics do not convey any meaningful message until we make sense out of them. The analysis and interpretation process thus becomes an integral part of indicator development. In the social science community, which largely followed the tradition of 'positivism', there was a strong professional conviction of remaining 'value-free' to maintain the scientific rigour of research by eliminating bias and prejudices as far as possible. The engagement of researchers in any pragmatic, either technical or political, project can be seen as falling into the trap of 'dual citizenship' (Berger and Kellner 1982: 136), which will ultimately compromise scientific integrity. Many social scientists prefer to take a healthy stance of being 'unattached intellectuals' (Mannheim 1936) who will then have a free hand to be critical of public policy (Hughes 1991). Those who have conducted policy-oriented research will no doubt have much sympathy with such a concern. The principle of scientific neutrality is a useful guide to strive for when carrying out data collection and analytical work. It is, however, inevitable that an analyst will bring his/her own personal values, knowledge and skills into the analysis. This is the very nature of any policy analysts; what matters is that the assumptions and the rationale underpinning of the analysis are made explicit and that all relevant technical and methodological information is carefully documented for transparency and public scrutiny. However, this is not always properly implemented and it is easy to straddle the very fine drawing line.

The 'modelling' processes used to attribute national survey data to local authority wards in the 2000 Index of Multiple Deprivation (DETR 2000b) is a useful example to illustrate the problem. Due to the lack of data sources to provide information on housing conditions at the local area level, the research team resorted to using survey data to carry out estimates. The scale of unsatisfactory private sector housing stock was estimated by modelling 975 unfit dwellings in an overall sample of 12,131 dwellings across England in the 1996 English House Condition Survey. The only information given in the consultation documentation was that the age and built form of the local dwelling stock, the economic circumstances of the local population and national patterns of poor housing were used as predictors in the estimation model. There was, however, a lack of detail on the estimated number of unfit dwellings for each of the 8,414 wards and the precise methodology employed in the estimation model, such as the 'explained' variance of the model and the residuals from it, so as to allow cross-checking and validation. The reason for not going into great length may not be related to the intention of concealing technical details, but possibly for the purpose of keeping the consultation document short and simple. However, this did cause concerns and speculation over the technical methods used.

In addition, the very nature of indicator research makes the prospect of drawing a defining line difficult. Since there is not a single perfect approach to developing a set of indicators, political choices have to be made even when deciding which indicators are to be included and what types of weightings to be used throughout the process. It is inevitable that open discussion and brainstorming sessions between the researcher and the policy client will take place. It is through this kind of debate, dialogue and discussion that the issues and concepts to be addressed are clarified and redefined (which echoes the point made by Ragin) to sharpen the relevance of the intelligence for policy-making. To make things more complicated, the policy client is not necessary equivalent to a single policy interest. In many cases, there are multiple and sometimes contesting interests in a particular set of indicators – the classic case is the development of deprivation indicators (further discussed in Chapter 4). In certain government departments, policy research is managed by officers who themselves have a strong research background. For instance, social researchers are embedded in four main analytical groups (i.e. Local and Regional Government Research Unit; Research, Analysis and Evaluation Division; Neighbourhood Research Unit; and Policy Directorates) in the Office of the Deputy Prime Minister (ODPM). As a researcher, I find the joint working relationship with most research officers in the ODPM very fruitful. There is a high level of intellectual interaction and exchange between the manager and the researcher, and the consequence is often a mutual learning process that enriches the input of the indicator project. Also, there is an explicit understanding of academic freedom and professionalism in the working relationship. It is, however, interesting to note that these research managers tend to work for an in-house policy client and they are the middlemen who broker the supply and demand of policy intelligence. The discourses of the nature and purpose of a particular indicator set obviously go beyond a two-way communication channel under such an institutional culture. In an ideal laboratory setting, one would like to separate the ivory tower from the political gutter during the clinical research process. In reality the interface between research and politics is very close when developing policy indicators. It is also arguably desirable to have the interaction and debate with policy-makers rather than cutting them off to avoid the risk of rendering academic independence. This means that we have to resort to the professionalism of both the researcher and the policy client to strike a balancing act.

Having accepted that it is possible to have dual citizenship, there is a further turn in the debate over the role of social scientists in the process of indicator development. Two alternative views were recently put forward by Innes and Booher (2000) and Sawicki (2002). They have very different perspectives over the value and function of indicators in the decision-making process (see Wong 2002b, 2003). Innes and Booher's argument very much focused on the political dimension of the indicator development process. They emphasised the importance of the user engagement process as a way of social learning to achieve consensus building. They were more concerned with the embeddedness of indicators in the decision-making process rather than the design and the technical substance of indicators. They argued that it was the former that contributes most to the decision-making process. Hence, they saw experts and social scientists as having an important facilitating role to play in consensus building. Sawicki, however, clearly advocated the role played by social scientists from the methodological perspective, or in his own words 'a rational paradigm' (2002: 25). While acknowledging the procedural arguments made by Innes and Booher, he raised doubts over the costs involved and the value of the consensus-building approach in influencing public policy. He also voiced concern that most community indicators projects were procedural rather than substantive in nature. This refreshes the memory of criticism made to the rational-comprehensive approach of planning as 'contentless' and 'contextless' (Thomas 1982), which contributed very little to our understanding of the substantive theory of planning. I suppose the concern is whether the emphasis on social learning and consensus building of indicators will overshadow the need of carrying out substantive analysis of social issues to inform policy decisions.

It is undeniably true that many well-designed indicators are very often ignored by policy-makers and never get near to influencing policy-making. As Pinder pointed out, promising research may be ignored because 'its message is unwelcome, because political circumstances or administrative personnel have changed, or simply because it comes too late' (Pinder 1980: 8). In a review of the Local Government Research Programme, the findings suggest that, in the context of a very fast-moving policy environment, policy-makers tend to consult experts and those with relevant experience to make quick responses and the efficacy of research is dependent on its timing in relation to the life cycle of a policy or an administration (WSA 1999: Annex E.5). It is then interesting to ask who are these experts and where do they acquire the knowledge? Many government think tanks are academics who have no doubt developed their expert knowledge through cumulative research efforts over the years. And as Blackman (1995: 192) commented, 'research does not appear to be practically useful. It refines the definition of problems and provides partial solutions but can leave policy-makers with more questions than they start with.' This very much echoes the policy enlightenment role of research as suggested by Weiss (1995). It is quite clear that research is only one form of policy knowledge and its influence is circumstantial to a number of factors. It is, nevertheless, reasonable to argue that just because research studies are not widely used does not make the effort put in the research redundant, as one can never tell when a piece of research will be used and its influence may not become explicit for years to come. More importantly, prudently conducted research will not stifle or distort discussion, although the analysis may not be immediately employed in the decision-making process. Nonetheless, a badly designed set of indicators could potentially cause tremendous damage to public debate. The research of Weiss and Bucuvalas (1980) reveals some surprising findings for the common belief that technical quality does not matter to policy-makers. They found that research studies conducted with high technical proficiency, and with critical, innovative and refreshing ideas, were more highly regarded by politicians and decision-makers.

It is explicit that researchers have a central role to play to advance the methodological and technical dimension of indicators. In this aspect, Sawicki's sentiments are shared. However, having said that, after trying to create reliable and valid indicators over the last ten years, I have also come to the realisation that many of the methodological and technical problems encountered in the development of indicators can only be resolved incrementally. The major stumbling blocks, such as the absence of clearly defined concepts, the lack of well-established causal theories to guide the selection of indicators and the absence of appropriate data at appropriate spatial scales, are unlikely to be removed in the very near future. The concern is, then, to focus on how we can communicate with policy-makers and users to help them understand these issues: to understand that there is a trade-off between using one approach from the other. There is thus a need to appreciate that the user engagement process has become more important, not so much to build up consensus, but more in its opportunities to extend an appreciation of the pitfalls and usability of indicators and to change more casual attitudes towards indicator use and data collection practice. Building on these standpoints, one would argue that both the rational paradigm and the communicative, social learning approach towards indicator development could and should co-exist. It is, however, my personal view that the adherence to methodological and analytical rigour and excellence should be considered as the necessary conditions of good indicator research, and a more inclusive and communicative approach to public engagement as the sufficient conditions for successful decision-making, and in that order.

INDICATORS IN USE: SOME KEY ISSUES

Since the demise of the social indicators movement, research interest in policy usage of indicators has dwindled. This is especially problematic in Britain as there is a dearth of comprehensive research on the use of research and intelligence by policy-makers (Stewart 1995b; Williams 1980). Although interest in using indicators to inform policy decisions has been apparent in the urban regeneration and environmental management fields since the late 1980s, there has been a lack of attention to the policy context and organisational approaches to the *utilisation* of indicators. Many longstanding issues of social indicators are, nonetheless, still relevant to the current discussion of the ways indicators are used. Three sets of issues concerning the use of indicators are raised here.

INSTITUTIONALISATION OF INDICATORS

In order to ensure the policy usefulness of indicators, several researchers have advocated the need to institutionalise indicators (Caplan and Barton 1978; Carley 1981; Innes 1990). Innes (1990: 232) defines institutionalisation as the setting up of routine procedures and practices to enhance the continuing existence of an indicator and to legitimise the method and concept of the measure. The purpose of institutionalisation is to produce and accept the measures, regardless of what they show, to avoid any opportunistic use or non-utilisation of indicators by policy-makers to meet with their particular ends. Institutionalisation can, however, create some significant drawbacks. Once the procedures and methods of measurement become formalised, it is very difficult to alter or replace them, even though flaws and unsatisfactory measures are found. Due to such inflexibility, the arrangements inherent in institutionalisation have to be carefully worked out to take into account of methodological as well as political and bureaucratic factors (Carley 1981; Innes 1990). The issue of institutionalisation has been a concern of British civil servants and politicians, as they have to set out the framework of measurement, as well as tying in the indicators with specific policy initiatives. The failure to institutionalise indicators, on the one hand, will inevitably create an impression that information is only used as a ritual dance; too much institutionalisation, on the other, will not fit with the promise of giving more flexibility to local policy development. It is thus important to examine how central government handles these dilemmas in the light of its recent approach to indicators usage (see Chapter 3).

THE ROLE OF DATA AGENCIES

The use of indictors inevitably links to the performance of data producers who help shape the roles of indicators in society through the tasks of data collection, measurement and analysis (Bauer 1966). Indicators as policy tools are thus subjected to the interpretation of different agencies' own perspective towards the phenomena in question (e.g. Innes 1990; MacRae 1985; Wong 1995). In order to avoid such bias, there is a suggestion (Innes 1990) that indicators should be produced by professional statistical agencies that have a strong awareness of policy issues, without having direct responsibility for them. MacRae (1985: 299) foresaw that a wider range of non-governmental information sources provided by 'information brokers', such as the private sector and other informal expert or technical groups, were increasingly playing an important role in shaping information systems. He argued that the independence of these information brokers, perhaps with aid from government, was more suitable for experimental and temporary statistics. The Royal Statistical Society (RSS 1995, 1999a) in Britain has long expressed concerns over the integrity of official statistics. The issues raised by Innes and MacRae highlight the importance of assessing the independence and integrity of the current operational framework of National Statistics (was the UK Government Statistical Service until 2000), and its inclusiveness in providing all statistics of public interest at different spatial levels. These matters will be further explored in Chapter 4.

STANDARDISATION VERSUS CONSENSUS BUILDING

There is a strong argument for regularising the methods and concepts of measurement (Innes 1990) to prevent any haphazard adjustment or manipulation of data. However, the issue of standardisation of local and regional policy indicators is a complex one. There are conceivable advantages in having a standardised set of statistics on local programmes to allow comparison and benchmarking of performance and progress among localities (Clark 1973). Nevertheless, different local areas have their own distinct development paths and the use of standardised measures inevitably conceals such local diversity and uniqueness (MacRae 1985). Moreover, it is important not to overlook the practicality of integrating local information sources into a standardised series, as they tend to be administrative records compiled under different formats and definitions. Standardisation may also undermine the perception of local citizens and policy-makers of what is the relevant measure to their specific local circumstances (Carley 1981). Innes (1990), therefore, called for an interactive model of knowledge development. This model encourages knowledge providers to improve public discourse and debate on the concepts, methodological approaches and usage of indicators. It, nevertheless, raises the worry that the process of consensus building may be biased towards those social groups who happen to participate (Innes 1990), thus reinforcing the political status quo to create information of 'lowest-common-denominator' values (MacRae 1985: 72). The contentious relationship between standardisation and local democracy over the development of indicators poses another major challenge to British policy-makers. The sensitive balance between these two sets of issues tends to be epitomised in the development of deprivation and sustainability indicators (see discussion in Part III of the book).

CHAPTER 3

CHANGING ETHOS OF INDICATOR USAGE

As briefly introduced in Chapter 1, the use of indicators in Britain has been closely related to the twists and turns in urban and regional policies, and resource allocation frameworks adopted by successive governments over the last twenty years. The changing political ideology and government ethos over public expenditure and policy monitoring have shaped both the methodology and the usage of indicators. The dramatic shifts are most notable in the pragmatic approach that has evolved around the usage of deprivation indices at both the national and local level. The discussion in this chapter aims to deal with the first three trends identified in Chapter 1 by providing a historical account of how the policy and political agenda moulds and shapes the attitude and ethos of indicator usage. It then explores the response of local policy-makers, in terms of attitude and capacity, towards the pressing requirements of using indicators via a number of recent empirical studies.

THE FORMULAIC CULTURE OF FUNDING ALLOCATIONS

There has been a chequered history of using indicators such as employment, unemployment and population changes for urban and regional policy targeting and evaluation. These measures were employed in the 1950s to identify the differences between Development Areas and non-assisted areas. Employment change, however, continued to be used as a key indicator to monitor urban and regional policies from the 1960s to the 1980s. Hughes (1991) provided a historical account of the evolution of the policy-monitoring culture in relation to the wider socio-economic contexts against which urban and regional policies were operated. He pointed out that, with the strengthening of regional policy in the 1960s, employment change and costs of industrial movement were the key measures of regional economic performance. He then suggested that, by the 1980s, following a long period of industrial and employment decline, the measure of policy performance on initiatives such as Urban Development Grant and Enterprise Zones shifted to focus on policy outputs such as the estimate of 'net' job creation, rather than the broader policy outcome of employment change. This pragmatic shift reflects the neo-liberalism enterprise culture of the government as the focus was on the costs and outputs of the policy and their incidence. There was a distinctive lack of concern for a broader focus of trend analysis, and the uncertainty surrounding the wider environment that has to be accounted for.

The widespread public debate on indicator measurement and usage, however, came later, after the publication of the Department of the Environment's 1981 Deprivation Index. In 1983 the Inner Cities' Directorate of the Department of the Environment (DoE 1983) published a paper *Urban Deprivation*, which provided analysis to assess the relative levels of deprivation in English local authorities. The 1981 Deprivation Index consisted of eight indicators (see Table 3.1) from the 1981 Population Census. Deprivation profiles of local authorities were developed by ranking their percentage scores on these eight indicators. Local authorities scored within the worst fifty in England on at least two of the indicators were deemed as the most deprived areas. In addition, six of the indicators (mortality rate and population change were excluded due to the lack of data at small-area level) were extracted at the Census enumeration district level. The indicators were standardised to derive four composite indices by varying the weightings attributed to the indicators (the formulae of the indices are given in Table 3.1).

Table 3.1 DoE's 1981 Deprivation Index

Key Indicators

- % unemployed persons
- % overcrowded households
- % single parent households
- % households lacking exclusive use of basic amenities
- % pensioners living alone
- % population change
- standardised mortality rate
- % residents living in households where the head of household was born in New Commonwealth or Pakistan

Four Indices

- Basic Index = 2 * (unemployment Z-score) + 1 * (overcrowding + amenities + ethnic + single parents + pensioners Z-scores)
- Economic Index = 4 * (unemployment Z-score) + 1 * (overcrowding + amenities + ethnic + single parents + pensioners Z-scores)
- Housing Index = 2 * (unemployment Z-score) + 2 * (overcrowding + amenities Z-scores) + 1 * (ethnic + single parents + pensioners Z-scores)
- Social Index = 2 * (unemployment Z-score) + 2 * (single parents + pensioners Z-scores) + 1 * (overcrowding + amenities + ethnic Z-scores)

Note: The indices were created by applying different weighting schemes to the standardised scores (z-scores) of the indicators.

The main reasons to develop the Deprivation Index were to inform the eligibility of local authorities to receive various physical regeneration grants under the Urban Programme and to provide background information to inform urban policy and to guide public expenditure at the local level (DoE 1993: para. 2). However, the widespread concern and debate over the 1981 Deprivation Index only came after the publication of the *Paying for Local Government* Green Paper (HM Government 1986) in 1986. The Green Paper proposed to allocate central government resources on the basis of the scale and nature of inner-city 'needs'. It then proposed to use a simplified and more stable assessment method of need to guide grant allocation and to reflect the differential costs involved in providing services by different local authorities (Flynn 1986). The DoE's deprivation indices were subsequently reproduced in the Audit Commission's Local Authority Profile and were used to allocate rate support grant to local authorities.

A technical exercise of index creation had turned into serious political debate once resources were found attached to the Index. Debate was then started on the conceptualisation of needs and deprivation, the choice of individual indicators, the methods and effects of statistical transformations, and the weighting schemes applied to develop the aggregate index. The comments made on the methodology and the choice of indicators by local authorities were all valid and genuine; it is however interesting to note that those authorities that put in a lot of effort to analyse the index were those that missed out from the top-ten most deprived areas, such as Liverpool (Flynn 1986; Hayes 1986) and Newcastle upon Tyne (Newcastle upon Tyne City Council 1986). This was partly because the top-ten most deprived areas according to the 1981 Deprivation Index were all London Boroughs, which would benefit most under the new rate support grant allocation regime. This top-down approach adopted by central government to resource allocation, mainly via consistent and systematic analyses of a variety of chosen quantitative indices to measure the relative level of local needs, has become both a technical as well as a political exercise. At the height of the Urban Programme, according to Cullingworth and Nadin (2002: 297), over 10,000 projects, costing £236 million (1992–3 figure), were funded each year in the fifty-seven Urban Programme Areas, as identified by the 1981 Index. This means that a significant amount of public resources are changing hands due to one single technical index. It thus becomes very important that such an index should be subject to more rigorous public debate and scrutiny. This is the main reason that the DoE and its successors' deprivation indices have continued to be a contentious issue of public debate. Another key concern surrounding the debate is the fitness of an index designed for a specific purpose, which is then applied to other policy uses without critically examining its limitations. Chapter 8 will provide further discussion on the methodological development and policy usage of deprivation indices.

Besides the Deprivation Index, the top-down funding allocation framework was also found in other policy areas. Annual capital allocations for housing investment by local authorities and the Housing Corporation are wholly based on indices of relative needs between different English regions. The formulaic allocation has been based on the 'Generalised Needs Index' for local authorities and the 'Housing Needs Index' for the Housing Corporation. In the 1980s, Census indicators were used to derive both indices to measure housing needs (Housing Corporation 1988). Due to the cyclical obsolescence problem of Census-based indicators, both indices have shifted to basing on data from a variety of sources in recent years. Since the late 1990s, the indicators were derived from the annual Housing Investment Programme returns, the annual Survey of English Housing, the five-yearly English House Conditions Survey and the Population Census (DETR 2000e). For regional policy, there has also been a long tradition of using unemployment rates, migration levels and industrial growth statistics by the Department of Trade and Industry (DTI *et al.* 1993, 1999) to map 'Assisted Areas' in order to determine eligibility for regional grant assistance to businesses to promote investment and innovation. The most recent review of Assisted Areas was carried out in 1999 (DTI *et al.* 1999) in response to new European guidelines on regional aid. During the three-month public consultation period, a wide variety of criteria were proposed for determining the new Assisted Areas. Labour market indicators, particularly unemployment, received the most support. Wards were the most widely supported unit of geography to form the basis of the map. The methodology finally adopted to devise the so-called 'Regional Selective Assistance Areas' was based on four indicators, namely, employment participation rates, residence-based unemployment rates, workforce-based unemployment rates and local dependence on manufacturing. These indicators were collected for groupings of local authority wards to identify areas with significant labour market weaknesses.

NEO-LIBERALISM AND COMPETITIVE CULTURE

As identified by Hambleton and Thomas (1995), one of the distinctive features of the urban policy since 1979 has been a strong desire to involve the private sector in the policy-making process. This approach to urban policy was very much part of the government's neo-liberal ideology that sought to replace much of the socialist welfare state with market mechanisms. The emergence of a global economy where there is no spatial constraint on the flow of factors of production, commodities, investment finance and information poses an increasing threat of international competition for capital investment and market share, and leads to dramatic shifts in the location of economic growth (Mair 1993). Hence, there was a belief within the Conservative government that it was up to localities and regions to promote their development, but not for government to redistribute jobs across the country.

The market-led enterprise ethos was injected into a whole array of regeneration initiatives by giving local authorities a more strategic role to draw up programmes of action to bid for regeneration funding (Hambleton and Thomas 1995). The allocation of regeneration resources through a competitive bidding process had started with the introduction of Estate Action in 1986 and City Challenge in 1991. This funding game was subsequently extended in the 1990s to a further suite of programmes including the Single Regeneration Budget, Regional Challenge, Local Challenge and Capital Challenge. Under the new competitive bidding arrangements, central government was less inclined to take a centralised overview of the relative social, economic and housing needs of different localities. Instead, there had been a shift towards a neo-liberal philosophy in which resources were allocated to local areas through the operation of quasi-market forces. This open competition for development funds had provided extra impetus to local coalitions and regeneration partnerships to make more flexible use of statistical information to advocate their case and target programme areas in their bidding documents. One example to illustrate this is again through the use of the official deprivation index (see Wong 2000).

The DoE commissioned researchers at Newcastle (Coombes et al. 1995) and Manchester Universities (Robson et al. 1995) to review and construct a deprivation index, which was subsequently known as the 'Index of Local Conditions' (ILC). The original expectation was that ILC would replace the former 1981 Index of Deprivation and be used by the DoE to guide the allocation of regeneration resources. However, in reality, while the bidding guidance gave prominence to ILC, it was seen as just one of the most important sources of information with which bidders might support their case for funding. The 1995 *Challenge Fund Bidding Guidance* to the Single Regeneration Budget (the SRB came into effect in April 1994 by pulling together virtually all regeneration resources from twenty different programmes to allow a more coherent approach to regeneration) stated that:

> Bidders should refer to relevant background statistical and other information which may be relevant to the bid. For example, in framing bids, particularly those which mainly aim at meeting needs, bidders *may* draw on information about local conditions in the 1991 ILC, and other data, including Employment Service information on labour market needs and gaps.
>
> (DoE 1995: para. 9).

This statement announced the Conservative government's intention to, at least partially, de-standardise and de-institutionalise the use of indicators in the allocation of urban regeneration resources. Although a standardised ILC was made available, it was no longer a compulsory requirement to use this set of information in policy targeting. On the one hand, such discretion brought plenty of scope for local policy-makers to make advantageous use of other statistics in advocating their cases. On the other hand, this change of approach was necessary as the design of the ILC as a *deprivation* index does not provide an adequate framework to address the entrepreneurial dynamics of localities for subsequent utilisation in a programme that was *competitive* rather than purely needs based. The reincarnation of the deprivation index into a 'Local Conditions' index only highlights the tension between the shifting political agenda and the original technical design of the index. There was, however, an interesting turn. Just not long before the election of the Labour government in March 1997, the DoE published a revised SRB bidding guidance. The new guidance adopted a much stronger tone in encouraging bidders to make use of the ILC and other relevant statistics to demonstrate the nature and extent of local needs. The 1997 *Challenge Fund Bidding Guidance* commented that 'bidders *should* draw on information about local conditions in the 1991 ILC' (DoE 1997: para. 10). This change of tone from 'may' to 'should' in the guidance indicates the re-institutionalisation of the ILC as a necessary part of submitted bids. The prominence of the ILC as a grant allocation tool was further reinforced in the July 1997 SRB Supplementary Guidance (DETR 1997a) and the November document on the allocation of £1.3 billion regeneration funds (DETR 1997b).

With the tightening of public expenditure, central government also began scrutinising public programmes and monitoring the effectiveness of individual policy activities more closely. Improved assessment and evaluation was required to co-ordinate different agencies involved in carrying out local development in order to avoid confusion over targeted areas. Spatial targeting and co-ordination of European aid also became more important, with the move towards Integrated Development Operation Programmes (CEC 1991). This led to the whole area of policy activity on performance measurement to demonstrate value for money. As admitted in a recent European Commission Working Paper (EC 2000a: 3), the monitoring, control and evaluation procedures of the Structural Funds had been largely financial monitoring. In reviewing the history of urban policy evaluation, Cameron (1990) commented that the Thatcher government made significant progress in urban policy evaluation, though with a tendency of using a restrictive number of performance measures such as job creation or retention rather than the broader impact and outcome of the policy. He further suggested that:

What is missing from this analysis so far is a series of indicators of the underlying 'health' of the economy (and, indeed, of the environment and of socio-economic well-being). Put in another way, we should devise indicators of beneficial directions which are themselves connected to beneficial processes in the local area. . . . A monitoring of policy outcomes measured at these deeper levels is a necessary component of defining and evaluating programme goals.

(Cameron 1990: 490–1)

This criticism of policy monitoring focusing on the development of intermediate output measures (such as hectares of derelict land improved and number of training courses provided) rather than the impact and effectiveness of regeneration activities in meeting policy goals was widely endorsed by other commentators (Burton and Boddy 1995; National Audit Office 1990). With the pressure imposed by central government, the auditing and performance ethos was also filtering to the management of service delivery in local authorities, though the extent to which different authorities had developed performance indicators varied considerably (Fenwick 1992). Again, the missing link between inputs (e.g. expenditure per head), outputs (e.g. number of new houses built) and outcomes/impacts (the outcomes for different groups or areas) of policy performance was seen as an issue of concern (Skelcher 1982: 50).

NEW LABOUR AND DE-STANDARDISATION OF STATISTICS

Despite changes in the central administration, the market-driven ideology continues to underpin the policy delivery of Tony Blair's Labour government. There is a continuous trend of shifting data collection responsibility from the centre to the local, but under strong central guidance. With regard to regeneration funding allocation, political signals from the then newly elected government suggested that the principle of competitive-allocated resources to coalitions led by local authorities was likely to remain in place. However, there was a socialist resonance of taking a greater account of *needs* when determining such resource allocations. The updated Index of Local Conditions was thus renamed as the '1998 Index of Local Deprivation' (DETR 1998b). These changes symbolise the intention of the government to provide a more standardised approach towards the use of statistical information in framing bids, and the focus of the Single Regeneration Budget programme has been reshaped to pay more attention to issues of deprivation and social needs. Nevertheless, there is still scope for local policy-makers to use other information to reflect their local circumstances in the bid. This means the final judgement of funding bids remains both political and unpredictable, and local actors remain confused about the role played by the Index and other statistical information in the bidding process.

The competitive ethos continues to be evident in the housing investment allocation. While the inter-regional allocation of housing capital is carried out on a purely formulaic basis, the approach towards intra-regional is another story (DETR 2000e). About half of the regional resources is distributed to local areas on the relative needs indices mentioned earlier, while the other half is at the discretion of the government (subject to the advice received from the Government Office for the Region and the Housing Corporation Regional Office). However, from the financial year 2002/3 onwards, the allocation of Housing Corporation investment has shifted away totally from a formulaic basis to full discretionary allocations. As expected, the discretionary allocation is carried out on a semi-competitive basis by making assessments on the relative performance and housing strategies of local authorities and registered social landlords within the region. This implies that a suite of performance indicators would still be required in the resource bidding process.

The trend of de-standardisation and de-institutionalisation of indicators was also introduced in land use planning. The idea of using indicators in regional planning was proposed in the *Future of Regional Planning Guidance* (RPG) consultation document (DETR 1998c) and the *Planning for the Communities of the Future* (PC) White Paper (DETR 1998d). The RPG document called for the use of indicators such as traffic levels and air quality to assess the achievement of objectives and issues set out in RPGs (DETR 1998c: para. 6.12). Further elaboration was made in the PC White Paper of the need to have housing provision indicators such as house and land prices, housing standards and local housing needs in RPGs (DETR 1998d: para. 31). Although there was an intention to use indicators in policy targeting and impact monitoring, these ideas were still very vague and general. After much criticism from the policy circles, a detailed report was produced to identify relevant data sources and relevant indicators to be employed to monitor housing provision (DETR 2000c). Some two years later, the Office of the Deputy Prime Minister (ODPM 2002a) issued good-practice guidance on targets and indicators for monitoring RPGs. The proposed monitoring approach from the ODPM is to link policy targets and output indicators with policy objectives. Further discussion on this will be made in Chapter 10.

The government also departed from its longstanding practice and opted for the de-standardisation and de-institutionalisation of household projection figures from local housing provision policy. The PC White Paper stated clearly that these projections figures 'were for guidance rather than, in effect, prescriptive' (DETR 1998d: para. 27). It will, therefore, be the task of local authorities to put forward their individual cases to justify their housing allocation policies. This sea change in government approach does signify a positive, democratic move for local and regional stakeholders to debate the issues relating to their local housing needs (Baker and Wong 1997). While welcoming this less mechanistic approach towards local planning issues, the key task will be to make sure that all interested parties are able to make their views known, and that there is a transparent process where the final jurisdiction is made. Another interesting turn to the matter is that, without prior consultation, the ODPM published the *Sustainable Communities* (ODPM 2003a) document. The Deputy Prime Minister called it an 'Action Programme' in his foreword. This action programme sets out the government's housing development proposals for a number of identified growth areas in the Thames Gateway, Milton Keynes, Ashford and London–Stansted–Cambridge. As stated in the document: 'Together with regional planning bodies and local authorities we will translate the development proposals for the growth areas into revisions of regional planning guidance so that they set out agreed levels of housing provision in the growth areas' (ODPM 2003a: para. 5.2).

Some serious concerns have been voiced over the scale of these proposed growth areas, especially when some do not have any existing infrastructure and transport links in place and some may consume excessive demand of water and other environmental resources (Morris 2003). While the document provides figures and charts to support the programme of action, there was no prior consultation or discussion with policy-makers at local and regional levels. It is interesting to find that, by de-institutionalising the household projection figures, the government can make the final jurisdiction of the housing numbers by asking regional planners and local authorities to revise the RPG. One may argue that this is the classic case where technical rationality and political rationality get mixed up, and it is difficult to attribute the exact rationale that underpins the policy decision.

MODERNISATION AND EVIDENCE-BASED POLICY-MAKING

The publication of the *Modernising Government* White Paper (HM Government 1999b) marked the government's strong commitment to performance measures and evidence-based policy ethos. The White Paper sets out the government's agenda on modernising the way policies and programmes are devised, the manner of public service delivery, and the way government functions are performed (HM Government 1999b: para. 4). The document is fuelled with buzzwords of performance, targets, objectives, audit and measures. In order to improve public service delivery and performance, Public Service Agreements and Best Value were introduced in the White Paper. Public Service Agreements are targets and measures set for all public bodies for improving public services with the aim to shift policy focus from inputs to the outcomes that matter. Best Value, however, is a locally defined monitoring system of service delivery and is underpinned by performance measures and independent inspection and audit. The duty for local government to secure Best Value was subsequently set out in the legislative framework of Local Government Act 1999 (DETR 1999d). A consultation paper on the performance indicators used to monitor Best Value was also published (DETR 1999c). The Best Value performance management framework has to be consistent with the existing and emerging Public Service Agreements between Government Departments and the Treasury. Since then, different sets of performance targets and measures were introduced in different areas of public service delivery. This was best summed up in an Audit Commission Management Paper *On Target*:

> The introduction of Best Value in local government, the Performance Assessment Framework in the National Health Service and in social services, the NHS Wales Performance Management Framework, and Public Service Agreements in central government has emphasised the importance of having good performance indicators as part of performance management in the public sector. To complement the national set of Performance Indicators, public sector organisations are expected to set their own performance indicators and targets, and to monitor and publish their performance.
>
> (Audit Commission 2000: 5)

The emphasis on the evidence base to underpin government policies was built upon the experience of the National Health Service (NHS). The term 'evidence-based practice' was first used in the NHS to describe the use of research evidence in policy, management and practice decisions (Blackman 1998: 56). The expectation of the government of rolling out this practice is that there will be 'more new ideas, more willingness to question inherited ways of doing things, better use of evidence and research in policy making and better focus on policies that will deliver long-term goals' (HM Government 1999b: para. 6). There is a belief that evidence is required to support policy choices and to generate public debate, especially in relation to the key factors and the way government policy affects outcomes (*Adding It Up* website, 2003). There is also a stronger emphasis on the importance of having robust analysis to underpin policy decisions. In order to kick-start a culture shift in policy-making, the Performance and Innovation Unit (PIU) of the Cabinet Office published the *Adding It Up* report (PIU 2000a) to set out a comprehensive programme for creating the conditions in which rigorous analysis is routinely delivered and demanded for policy-making. A range of initiatives has then been introduced to encourage more adequate use of research evidence in policy formulation. These initiatives include the launch of the *Evidence for Policy Choice* website in June 2002, a cross-departmental 'Evidence-Based Policy Fund', a range of high-level seminars for civil servants and a short-term academic placement scheme. The Evidence-Based Policy Fund is administered by the Treasury and has been allocated a £4 million seed corn fund to promote the supply of research and analysis for crosscutting policy, and to strengthen the links between universities, research institutes and the government by funding applied research on a number of priority topics identified by the government.

These latest moves highlight that a strong culture shift towards research-based, evidence-based policy-making is well underway. According to Sanderson *et al.* (2001), the need to secure Best Value serves as the strongest driving force to develop research and to integrate it to management in local government. However, sceptics such as Innes (2002: 102) describes this 'ready, fire, aim' practice as a pseudo-scientific approach to policy-making as well as a nice fantasy. The challenge to the rhetoric of providing proofs and facts to support policy decisions is largely impinged on the inherent tension between scientific rationality and complex political reality. Ironically, the introduction of an evidence-based policy regime is not based on any firm proof that there is a direct relationship between research and policy decisions (as discussed in Chapter 2). However, as argued by Weiss (1995), there is an enlightening role played by information and research in the policy-making process. Reading the lines on the *Adding It Up* website carefully, it is clear that 'policy choices' and 'debate' stimulated by the facts and evidence is what the government sees as important. The shift to an information-demanding policy regime is closely related to the government's belief in consumer culture. The public is the customer or client of public services and that they should have the right to know and to be informed of government performance. Hence, performance league tables of schools and hospitals are now published on government websites. In spite of much political pressure from schools and the teachers' union to remove these league tables, as they are subject to misinterpretation and stigmatism, these tables are widely welcomed by ordinary parents as they believe they have the right to know. The invention of Best Value indicators is just an extension of these league tables to name and shame those underperformers. Since resources for public service delivery are highly constrained, the idea is to use consumer demand as a lever to manage the supply of these services. This can be seen as a shift from a resource-led to a demand-led management regime.

Another important signal projected from the government's monitoring guidelines is the increasing emphasis on the longer-term horizon of outcome and impact measurement (DTLR 2002; HM Government 1999b; ODPM 2002a, 2003b, 2005b). This coincides with the latest guidelines issued by the

European Commission (2000a) over the monitoring of the new programme of structural assistance. The operational monitoring framework proposed includes the development of indicators to measure inputs, outputs, results (direct and immediate effects) and impacts (longer-term effects). There seems to be a realisation that the focus on monitoring outputs in the late 1980s and early 1990s failed to provide a clear picture of policy achievement and there is a need to expand the scope of evaluation to allow the monitoring of trends and changes (EC 2000a; SEU 2000). Hence, the concept of baselines and contextual indicators are frequently mentioned to allow more rigorous evaluation of progress and change (EC 2000a: 11; DTLR 2002; ODPM 2003b, 2005b). If the pertinence of evidence base embraces more rigorous analysis and a more positive approach of policy evaluation, it should be widely welcome in the decision-making process.

DISCRETIONARY LOCAL INTELLIGENCE

In spite of the fact that the government is the largest provider of information (see Chapter 4), it has increasingly favoured a less centralised approach towards the use of information in local policy decisions. The de-standardisation and de-institutionalisation of the use of deprivation indicators and household projections marks a watershed from its previous 'top-down' approach towards indicator usage. This more hands-off approach can be seen as a positive move towards democracy that provides opportunities for local stakeholders to identify their own problems and voice their concerns. However, flexibility comes with responsibility and the central–local shift can also be cynically viewed as a way to release central government from the political limelight when making unpopular local decisions. It may also reflect the weakening of government capability to go through the difficult process of achieving consensus to develop a transparent and standardised set of statistical tools to fuel public debate. It is also important for the government to issue more transparent guidance on how it will use part standardised and part non-standardised information in decision-making. This is likely to be especially contentious in respect of local housing provision decisions, following the proposal to downgrade the importance of national household projections figures and the subsequent introduction of the Sustainable Communities action programme to dictate the development blueprint across different parts of England.

The strong push towards the evidence-based policy approach of decision-making signifies the start of an information intensive governance regime. However, on closer reading, ambiguities and confusion over the rationale and purpose behind measurements are frequently found. In many cases, the development of indicators tends to be driven by *ad hoc* policy concerns. Having some sort of indicators seems to be a political panacea in many recently published documents. This issue will be further illustrated in Part III of the book when discussing the development of various policy indicator sets. There are two main reasons behind such confusion: the relationship between indicators and their ultimate policy use is not always clearly understood or defined, and there is usually a lack of any sound theoretical framework to guide the selection and interpretation of indicators. Hence, a compromised and unsatisfactory set of statistics will often be developed out of a common-sense, empirical approach that does not serve any particular policy purpose.

The decentralised, hands-off approach used by central government over the application of statistical information in regional and local policy-making has increasingly shifted the responsibility of information collection and usage to local policy-makers. Nevertheless, there is a strong tendency for Whitehall to issue good-practice guides on how to develop different types of indicators to monitor and evaluate policy. Furthermore, the government is ready to offer 'the carrot and the stick' by attaching funding resources to performance indicators. This creates a paradoxical situation of decentralisation and centralisation at the same time over information usage. It is, therefore, interesting to find out how local actors react to such a heavy-handed steering from the centre.

The reactive culture of local authorities towards central government pressure of integrating research in policy-making can be traced back as far as the 1960s. Deakin (1982) provided a very comprehensive and interesting historical review on research and policy-making in local government. The historical discussion here largely draws on his findings. Town planning has definitely played a key role in bringing research activities into local policy-making (see Deakin 1982; Donnison 1975). Following the publication of the *Future of Development Plans* report (PAG 1965), a new system of structure planning was in place in 1968. The preparation of structure plans closely followed the rational comprehensive approach of resorting to data collection and analysis. As Deakin observed, 'the day was not far off when no meeting of a planning committee would be complete without knowing references by members to modelling and modal splits' (Deakin 1982: 304). In 1968, the Seebohm Committee carried out a review on personal social services and came to the conclusion that research capacity was very important and the power to assemble and analyse information should be made available to all departments in local authorities. The reorganisation of local government in the early 1970s provided a catalyst for the development of corporate planning and policy analysis in many new local authorities. This was encouraged in the Bains Report (1972), and the 1972 Local Government Act provided the power for the expansion of research facilities in local government. Meanwhile, local government was bombarded with central government guidance on the way to conduct their affairs, mainly via the use of research and intelligence. The then Secretary of State for the Environment, Peter Walker, was pointed out as a leading figure behind the expansion of research activity during the 1970s. Walker was a firm believer in the rational model of planning by objectives, informed by research and data analysis.

After the rapid expansion of research activities in the mid-1970s, the mood towards research within local government was, however, less favourable by the end of the decade. This was in part related to losing faith over the performance of the reformed planning system. Under the cost-cutting culture of the Conservative government, central government was also more lukewarm towards the idea of policy through analysis. Hence, research activities in local government went through a phrase of downsizing and contraction as a result of cutbacks in the 1980s (Blackman 1995). More importantly, the abolition of the metropolitan counties in 1986 was found to be a nightmare to local government research as a lot of valuable information, computer systems and datasets were lost in the turmoil (Gilfoyle and Wong 1998). This was partly compounded by the problem that the key personnel who were involved in the day-to-day management and operation of information and research either left for other jobs or took early retirement. Davies (1996) thus voiced his concern over the potential losses of data following the latest round of local government reorganisation that started in April 1996.

With the emphasis on needs-based assessment, policy monitoring and evaluation, competitive bidding of public funds and performance measures, interest in research and information analysis was once again rejuvenated in local government in the 1990s (Blackman 1995). However, with the severe contraction of research capacity throughout the 1980s, there has been concern over the mismatch between the availability and requirement of research skills. Furthermore, the range and level of research activities conducted also varies dramatically from authority to authority. As Blackman observed, larger councils tended to have the capacity to undertake the greatest amount of research and to employ research officers directly. It was against this changing context of local government research activities that the Local Authorities Research and Intelligence Association (LARIA) commissioned the Centre of Urban Studies at Bristol University (Boddy and Snape 1995) to carry out a national study on the role of research in local government in 1995. After a national survey of local authorities and some in-depth case studies, the study concluded that 'research and information remain, however, a largely discretionary activity, themselves under threat from resource constraint and local government reorganisation' (Snape and Boddy 1996: 39). The study also found that nearly three-quarters of surveyed authorities felt that research made some contribution to policy-making and 17 per cent saw it as a key component. More importantly, nearly three-quarters of local authorities thought that the policy impact of research had increased over the past three years. The *ad hoc* use of research in local government is further compounded by the nature of some research studies, which tend to involve data collection and the production of simple tabulations and diagrams rather than in-depth analysis to yield policy intelligence (Blackman 1998).

The issues raised in the above studies very much echo the findings of an Economic and Research Council-funded project carried out by the author (Wong 2000). This research study explored the attitude and capacity of local policy-makers in the local economic development (LED) field towards the usage of indicators in two English regions. One of the key findings was that the audit culture and information-intensive monitoring approach adopted by central government and the European Union, to a large extent, motivated local actors to look for more relevant information and socio-economic indicators. Such information also provided a basis for practitioners to discuss the key issues with their partners in putting their case forward. However, in spite of the awareness of the pressing need to have statistical indicators, many LED organisations admitted that they were not well prepared to meet such requirements. Most of the interviewees agreed that information collection in their organisation was carried out in an *ad hoc* manner and a few of them were frank about the absence of any monitoring activities in their organisation at all. A lack of staff and financial resources were most frequently mentioned as reasons for their lack of capacity to have a proper evaluation or monitoring system in place. It was evident from the in-depth interviews that there were subtle differences between policy-makers in the two case study regions in terms of their attitude and approach used in data collection. Those in the Eastern Region tended to adopt a more haphazard approach, and were less inclined to collaborate with other organisations over data collection. Their lack of urgency has to be viewed in the light of their local context as the Eastern Region consisted of many affluent areas that were not eligible to apply for central government or European funding. The situation was significantly different in the more deprived North West, where practitioners were desperate to argue their case for all sorts of regeneration funding.

Another interesting finding from the study was the urge of seeking more information without specific policy purposes. At the first glance, it is encouraging to find that over three-quarters of the respondents in the survey found indicators useful to their work. However, when prompted for further explanation, only one-third of them envisaged using indicators in formulating policy, a fifth in comparison and monitoring, and even fewer in preparing bids for funding (see Table 3.2). The survey findings thus highlight concerns over local policy-makers' lack of a sense of direction over the use of indicators in supporting decision-making, in spite of the possible resource incentives from central government and the European Commission. The overall picture of information usage among the participant

Table 3.2 Potential application of LED indicators (survey response)

Use of indicators	North West (%)	Eastern Region (%)
Policy formulation	36.1	32.8
Comparison & monitoring	19.4	25.0
Prepare funding bids	12.5	12.5
Background information	11.1	9.4
Marketing & promotion	11.1	4.7

Source: Wong, 2000: 232.

organisations could be characterised as somewhat patchy and disorganised. This probably reflected the fact that there were significant hurdles that prevented policy-makers from applying quantitative indicators in their work, in spite of their vague, general interest in seeking such information.

The key obstacles encountered by policy-makers in accessing relevant LED information were found to be very similar to those confronted by academics (see Table 3.3). A lack of relevant information sources (such as property and business statistics, skill level of the unemployed, and income data) was most frequently mentioned by the practitioners. Even where certain information was available, a number of respondents criticised the quality of local statistics, such as the Census of Population and the Labour Force Survey, as unreliable and inconclusive. They also raised the concern of the lack of spatial breakdown of information. Many data sources, such as public expenditure on research and development, did not go below the regional level. Other than the disaggregated data from the Census of Population, many interviewees would like to see more flexibility in area specification and be able to access statistics at local authority district and, ideally, even at lower spatial scales. Since data linkage was seriously hampered by variations in the spatial scale of data compilation, the choice of appropriate building blocks was thus considered as very important. They also pointed out the problem of incomplete spatial coverage of some data sources, which created reliability problems in comparative analyses as they could not be compared on a like-for-like basis. The final identified problem was, however, a political one. There was a general lack of trust from local policy-makers towards the ways statistical information were compiled and used. Complaints were made on the lack of consultation from the government on the usage of statistical indices in resource allocation. Some participants, both in the survey and in-depth interviews, stressed the importance of having more qualitative information rather than just producing a league table of statistics. They were dissatisfied with the fact that statistics were frequently subjected to manipulation, and were very hostile towards the hard performance measures that only produced a partial story of regeneration and economic development.

Table 3.3 Problems in assessing LED-related information (survey response)

Problems in information access	North West (%)	Eastern Region (%)
Availability of information	15.3	18.8
Quality of information	1.4	4.7
Spatial scale	9.7	17.2
Spatial coverage	5.6	6.3
Updatedness of data	2.8	10.9
Organisation of data sources	6.9	4.7

The LED empirical study of the North West and Eastern Regions highlights the reaction of local actors towards the government's new, information-intensive, approach to decision-making. Whilst acknowledging the importance of having indicators to lobby for more resources and inform decision-making, many local policy-makers did not really have a clear idea of what indicators could be used for, and they were not well prepared to cope with the need to manage statistical information in a systematic manner. The findings show that something of a culture shift of using indicators in policy-making has been underway among local practitioners since the mid-1990s. However, the pace of change is somewhat slow to match that of central government. The obstacles are three-fold: first, there is a lack of confidence over the integrity of government statistics and their usage in policy decisions; second, some local actors are overwhelmed by the need for more information without a clear understanding of its usage; and, finally, the national statistical infrastructure does not produce adequate information to serve local policy needs (further discussed in chapters 4 and 5). These concerns are similar to those raised by Blackman (1998) and Boddy and Snape (1995).

EVIDENCE-BASED GOVERNANCE: MORE REACHING OUT

The empirical findings of the LED study discussed above were conducted during 1995–7. Since then, with the arrival of the Labour government, the evidence-based policy ethos has exerted further pressure on local policy-makers to make an effort to collect and analyse information on a systematic and regular basis. At the same time, the difficulties and unnecessary duplications of data collection are widely acknowledged in the *Better Information* report (SEU 2000). However, the observations made in the LED study in many ways are echoed by the findings of another major study (Percy-Smith *et al.* 2000) carried out for the Improvement and Development Agency to examine the way research is used to inform decision-making in local authorities in response to the local government modernisation agenda. By the time the study was conducted in 1998/9, research findings were found to be used to serve instrumental functions and to provide factual information in local government. The key driver of research in local government tends to come externally from central government through performance measures and competitive bidding of resources, and 80 per cent of authorities in the study mentioned that lack of resources was the major obstacle to research. Nevertheless, the top-down pressure from central government through the introduction of Best Value and evidence-based policy-making seems to provide some stimulus for local authorities to develop their research capacity.

My involvement in the development of a Town and City Indicators Database (TCID) for the Office of the Deputy Prime Minister (ODPM) in 2002/3, to monitor progress made towards achieving the urban renaissance visions of the Urban White Paper (DETR 2000d), has allowed me to revisit some of the issues identified earlier in the LED project. There is now a general understanding in government departments that they need to avoid putting the burden of extra data collection on other organisations and that there is a need to develop co-ordination and co-operation over data sharing and data collection. Hence, one of the basic premises of the TCID project was to make use of existing datasets in the public domain rather than to collect new data (ODPM 2002a). However, it is easier to respect the principle than to apply it as indicators are tailored for different purposes and it is impossible to go back to the 'one size fits all' approach. Sometimes an indicator may appear to be quite straightforward at the first glance, but its meaning and interpretation could change dramatically when a slightly different denominator is applied. Many of these issues are fully discussed in a report published by the ODPM (Wong et al. 2004). However, I would like to pick out a few interesting points from the in-depth interviews with key actors in six urban areas.

Through case study interviews, it is encouraging to find that many local authorities have made good progress in handling the information-intensive policy regime and some have come to realise the benefit of having high-quality information to facilitate partnership working and policy debate. It is, however, interesting to note that the two urban areas that are less inclined to collect data tend to be local authorities in more affluent locations. Due to the economic success in these areas and the relatively low level of social deprivation, there is no incentive for these areas to spend resources on developing their research capacity or mounting any major data compilation exercise to enhance funding bids or policy monitoring. The situation has not really changed since my earlier study in 1995/6. The capacity and aptitude of using indicators to inform policy-making is still much less developed in more affluent areas. This is simply because there is no external pressure or threat to do so. HenceS, research and intelligence in these areas is largely operating on the 1970s models that are heavily reliant on the delivery of town planners as data-collecting agents because policy monitoring is closely associated with statutory planning functions. Having said that, with the kicking-in of Best Value indicators, even these areas are now engaging in efforts to collect certain types of information, at least, to fulfil the requirements of the Audit Commission and the Treasury. On the contrary, for urban areas that have undergone major socio-economic restructuring and suffered from economic decline and social deprivation, there has been a strong emphasis and awareness on the importance of making use of indicators for policy and programme monitoring as well as developing performance measures. The problem remains regarding the difficulty of accessing reliable information at small-area level for local policy. Local actors are also overwhelmed with the demand for all sorts of performance and monitoring indicators. Data are collected by different partners and agencies at the local level without much co-ordination, which in turn leads to unnecessarily duplication and confusion. These problems are actually acknowledged in the Performance and Innovation Unit (PIU 2000b) report *Reaching Out: the Role of Central Government at Regional and Local Level*.

To conclude the discussion, the evidence collected so far strongly suggests that the development and use of indicators at the local and regional level is heavily reliant upon the guidance and supervision of central government. In spite of the fact that central government has increasingly favoured the use of indicators to inform decisions on urban and environmental planning issues, it is clear that some are more conceptually developed and embedded in the decision-making process than others. Those indicators that are taken more seriously by policy-makers tend to be those with a strong and clear link to public resource allocation. One clear example is the use of deprivation indices. In spite of the flaws and recent confusion over their policy usage, deprivation indicators have a higher degree of institutionalisation. This is partly due to their long history of development, and partly related to the fact that there are specific policy needs to make use of such information for resource allocation. Hence, there has been continuous effort to refine and update the indicators. Other examples include the recently introduced Best Value and Public Service Agreements indicators; their prominence as performance measures is closely related to the public funding allocation regime.

The new culture towards accountable public funding has, to some extent, already helped foster the political will towards policy monitoring and evaluation. It is explicit from the LED study that the ultimate stimulus of behavioural change comes from 'the carrot and the stick' approach of central government and the European Union to link indicators to funding allocation. The issues over the integrity of government statistics, the scope of having more disaggregated spatial information and the need to have a code of practice on all public statistics will no doubt continue to be debated. There is also a continuous need for the government to issue clearer policy frameworks to explain how indicators will be used to inform public funding allocation and policy decisions. The institutionalisation of certain indicators with funding initiatives will certainly facilitate a culture shift. However, this has to be carefully thought through, otherwise it will just distort policy attention and priority. There is also a conundrum here in that the decentralisation trend of information collection and use is very much driven by a strong guidance framework from the centre, through the carrot of funding allocation as well as the stick of resource curtailment. The question is whether an evidence-based policy culture can ever be fully embedded into local policy-making without financial control from central government to provide the impetus. This remains an interesting issue for future research.

CHAPTER 4

MANAGEMENT AND ORGANISATION OF NATIONAL STATISTICS

BETTER INFORMATION

For a long time, the issue of public information management has been sidelined and neglected in Britain. The knowledge of data sources and their strengths and limitations has become the expertise of a minority group of consultants, academics and government statisticians. It is, therefore, so refreshing to see that the *Better Information* report from the Social Exclusion Unit's Policy Action Team 18 confronted the problem:

> Many commentators were impressed by the number and quality of the statistics that back up the report [the 1988 report on deprived neighbourhoods]. They might have been surprised to know how difficult it was to pull the information together from a series of disparate and often incompatible sources.
>
> (SEU 2000: 7)

The Policy Action Team's comments on information management and organisation issues provided a comprehensive and authoritative review. The report made some strong but accurate criticisms of the difficulties and unnecessary duplication of data collection:

> This is not to say that information does not exist – somewhere. Government collects information about the people and the facilities in these [deprived] areas all the time. But much of this information remains hidden away in the computers and filing cabinets of the people who collected it, unused because its owners did not know how useful it might be for other services to have access to it. Sometimes the owners had never been asked to share it, because no one else knew they had the information. Sometimes it was not shared because someone thought wrongly that sharing statistics was illegal.
>
> (SEU 2000: 7)

In the face of an information-demanding policy regime of the British government, this chapter turns the attention to the organisation and management of information and assesses the adequacy and openness of the national statistical infrastructure to support policy decisions. It also explores the development in the provision of local and regional intelligence for decision-making.

STATISTICS: A MATTER OF TRUST AND QUALITY

Until very recently, the Government Statistical Service (GSS) was by far the largest provider of statistics and had the greatest concentration of statistical expertise in the United Kingdom. The GSS was a part-centralised, part-decentralised service that comprises the Office for National Statistics (ONS) and the statistics divisions of all major government departments (see HM Government 1998). The discussion here examines the structure and accountability of the GSS before the government's recent overhaul of national statistics through the publication of the *Statistics: a Matter of Trust* Green Paper (HM Government 1998) and the *Building Trust in Statistics* White Paper (HM Government 1999a), which led to the final introduction of the *Framework for National Statistics* reform in June 2000.

The ONS is an independent agency created in 1996 by the merger of two big collecting agencies: the Central Statistical Office and the Office of Population Censuses and Surveys. It aims to provide greater coherence and compatibility in compiling and maintaining a central database of key statistics, as well as the production of the periodic census of population. The remit and function of the ONS to provide 'an authoritative and impartial picture of society and a window on the work and performance of government' (ONS 1998: cover page), in many ways, resembles the characteristics of the professional statistical agency portrayed by Innes (1990). However, it is the Chancellor of the Exchequer who is accountable to parliament for the activities and resource allocations of the ONS. This is further complicated by the fact that individual government departments also produce statistics within the working and resource framework determined by their respective ministers. Hence, Bauer's (1966) concern that information could be biased by the political perspective of government departments remains an issue today.

The only checks and balances, before the publication of the Green Paper, came from the co-ordinating role performed by the Head of the GSS (who was also the Director of the ONS), and the Statistical Advisory Committee that provided advice on statistical issues and priorities. Under this framework, the integrity of statistics critically depended on the professionalism of those who were involved in the production of statistics. Unlike many other countries, there was no administrative statistical legislation in the UK to safeguard the integrity of the operational framework. There was also a lack of an independent body that determined the scope, quality and integrity of the statistical service

(RSS 1998). Hence, it was not a surprise to find lingering suspicions of official statistics, such as the measurement of unemployment, being politically manipulated (Bartholomew 1995; Beatty et al. 1997). The shortcomings of the current system were openly admitted in the *Statistics: a Matter of Trust* Green Paper.

The Green Paper explored the pros and cons of four options to improve the accountability of the system and to provide a governance framework of official statistics. These four models were:

strengthening existing arrangements,
establishment of a governing board with a non-executive chair,
establishment of an independent Statistics Commission, and
direct accountability to parliament.

The latter two models no doubt are more accountable; however, the cost incurred would also increase. The final arrangements were thus very likely to represent a trade-off between the objectives of integrity and the constraint of resources. As the Green Paper stated, these models were not mutually exclusive and the final recommendation could be a hybrid model combining some of these options together.

This move in reviewing the GSS signified a major step forward in building up the credibility of the national statistical infrastructure. This major consultation exercise was described by Melanie Johnson, the Economic Secretary to the Treasury, as 'the widest debate on official statistics for over 30 years' (HM Government 1999a: v). The integrity of statistics was very important, if they were to be widely used to inform policy decisions and stimulate public debate. The Royal Statistical Society (RSS), a learned society, however, believed that it would take time to achieve a culture shift and 'none of the options presented in the consultation document would, on their own, achieve the desired and necessary change' (Curnow 1998: 276). The RSS (1998) also commented on the rather narrow definition of national statistics covered by the proposals in the Green Paper, and it urged the government to extend the scope of national statistics to include statistics produced by all public sector bodies at all geographical levels. This issue is of particular importance to the development of spatially oriented planning indicators. Having some forms of quality control over public data collection practice, to enhance meaningful comparison of changes between periods of time and geographical areas, have long been urged by planners (Wong 1993; Worrall 1991). Despite the publication in 1995 of the *Official Statistics Code of Practice* by the Head of the GSS, which set out principles and practices to govern all official statistical work, the code did not necessarily ensure consistency in the definition and spatial scale of many data series. The calculation of economic activity rate in *Regional Trends* (e.g. ONS 1998) is a typical example, which can be expressed as either the population of working age (in Table 5.3) or people aged 16 and over (in Table 5.1). The situation gets worse, as the 2001 Census data are released on the basis of 16 to 74 years old. The lack of a single definition of indicators that is widely used by policy-makers can easily cause confusion in policy discourses and lead to meaningless comparisons.

The spatial comparability of government statistics has also been problematic. This is partly due to the fact that public data series tend to be published for administrative boundaries that are susceptible to political changes. Since the abolition of the metropolitan counties and the Greater London Council in 1986, many data series have been revised to release data for shire counties as well as the new metropolitan districts and London boroughs. Nevertheless, great variations are found in data publication practice. For example, only until recently, both the New Earnings Survey and the HM Land Registry's residential house price reports only published data for shire counties and London boroughs, but no statistics were provided for individual metropolitan districts except aggregated data for their abolished counties. Following the latest round of local government reorganisation in 1996, many new unitary authorities have been set up to replace the former two-tier structure. However, in contrast to the earlier situation, government departments and the ONS were not slow to get their act together to release data for these new unitary districts within a very short period of time. The volatility problem of administrative units is further exacerbated at the lower spatial levels such as wards. For the future, efforts are being made to move away from using administrative boundaries and to derive more data for micro areas, so as to produce data for accurately defined study areas. The recent development of Super Output Areas (built from clusters of adjacent unit postcodes) by the Neighbourhood Statistics Services will offer a more promising prospect of data aggregation. This will be further discussed in Chapter 5.

In order to overcome inconsistencies in the published data, users can sometimes use the raw data to make their own adjustments. For instance, house price data are now available at the postcode sector level from the HM Land Registry, value-added tax registration data can be obtained from the ONS and vehicle stock data can be purchased from the commercial selling agents of the Driver and Vehicle Licensing Agency. However, processing raw data into usable formats usually involves technical skills as well as data purchase costs. Due to the lack of resources, local authorities can be discouraged from using these valuable statistical sources. Following the trend of commercialisation of information (Coombes and Wong 1994), the role played by the private sector and other research institutions in analysing and disseminating policy-relevant statistics will increasingly be seen as a key factor influencing policy usage of information. It is also clear that the objective of having a more coherent practice of public data compilation is not an easy task. Nevertheless, the call by the RSS to extend the scope of quality control to all public sector statistics helps to set the agenda on the right path. Until more harmonisation in the definition of measurements and better data quality control are imposed on all public sector statistics, the objective of conducting meaningful analysis of temporal change and spatial distribution to aid policy design and monitoring will remain an unreachable ideal.

TOWARDS NATIONAL STATISTICS: TO BE OR NOT TO BE

Before the Labour Party was elected to government, they made a manifesto pledge of reforming national statistics to regain the trust and confidence of the public over the independence and openness of official statistics. The publication of the *Statistics: a Matter of Trust* Green Paper has raised expectation over the wholesale overhaul of the national statistical infrastructure. However, after the initial enthusiasm of publishing the Green Paper, the final publication of the White Paper *Building Trust in Statistics* was done in a somewhat hurried manner. This is evident from the immediate response made by the RSS that:

> The RSS ... welcomed the speedy publication of the green paper to build on the manifesto pledge. But that was in February 1998 and since then we have heard nothing until today when a White Paper *Building Trust in Statistics* has been published. There was no prior notice of this publication, a small restricted press conference and it has arrived only a matter of hours before a (rare) debate in Parliament on Official Statistics.
>
> (RSS website 1999a)

The White Paper aims to provide a platform to ensure public confidence in official statistics. It proposes a new framework for National Statistics with the premises to:

strengthen statistical priority-setting and responsiveness to all users;
ensure professional freedom in the operational production of statistical outputs;
ensure statistics are produced to high professional standards; and
provide greater transparency and accountability than current arrangements.

(HM Government 1999a: para. 2.1).

The major structural change brought by the White Paper is the establishment of an independent, non-executive Statistics Commission to advise ministers on priority setting and quality assurance matters, and to provide advice on the scope of National Statistics. The Government now also appoints a National Statistician, who has an overall responsibility to oversee the outputs of National Statistics as well as taking over the responsibilities of the Director of the ONS and the Head of GSS. Statistics continue to be collected at different government departments and the National Statistician has to perform a strong co-ordination role. Rather than directly accountable to parliament, similar to the past system, the National Statistician is accountable to the Chancellor of the Exchequer. The need to have a legislative framework was rejected in the White Paper on the ground that it would take considerable time to implement any new arrangement. Hence, the proposed changes were implemented on a non-statutory basis. In their formal response to the White Paper, the RSS highlighted the tension caused by the dual roles played by the Treasury, both as the statistics users and the parent department of the ONS (RSS website 1999b). Strong reservation was also expressed by the RSS over the missing opportunity of providing a legislative framework to ensure the independence of the statistical service. As the proposal stands, it is really a matter of trust from the public that the government will stand by its commitment made in the White Paper.

The other major change brought by the White Paper is the concept of National Statistics and its scope in relation to those datasets that fall under the umbrella of GSS. National Statistics refer to the preparation and production of statistics intended for public use and there will be a Code of Practice to provide quality assurance of the statistics released. It is, nevertheless, interesting to note at the onset of the reform the White Paper provided the preamble that not all statistics covered by the GSS would be included in the National Statistics framework: 'Since many of the statistics produced by the GSS are used for many purposes, to attempt to "kitemark" each statistics for every use made of it would raise significant practical difficulties' (HM Government 1999a: para. 3.3).

> The other key element to demonstrating integrity is to ensure that official statistics are produced without political interference and that that is clearly recognised by users. The new framework is designed to ensure that the professional responsibility for National Statistics is clearly separated from the responsibilities of Ministers. The National Statistician will set professional standards for National Statistics, including standards for release arrangements and quality assessment, and will have the authority to determine whether or not a particular statistical output meets National Statistics requirements.
>
> (HM Government 1999a: para. 3.5)

The selective and narrow definition of National Statistics has caused disquiet from the RSS that:

> there will be less accountability and coherence even than now: currently the Director of ONS has a role in relation to the GSS whereas the National Statistician and the Statistical Commission have not been assigned authority for statistics falling outside the scope of National Statistics.
>
> (RSS website 1999a)

This means that:

> many key statistics of most direct interest to the public, including hospital waiting list data, school and university league tables, class sizes, statistics relating to BSE, rail safety information, police force numbers, will be outside the arrangements unless Ministers determine otherwise.
>
> (RSS website 1999b: para. 10)

While the government is arguing for a more harmonised approach to statistics, or in its own words 'joined-up statistics' (HM Government 1999a: Appendix B.14), it seeks to achieve this by stepping backward and adopting a narrower definition of national statistics. This clearly countered the argument made during the consultation process to widen the definition of official statistics. This half-hearted reform of the national statistical infrastructure does cause concern over the quality of those datasets produced outside the National Statistics framework. For example, the 2000 Index of Deprivation and its successors, used to allocate regeneration grants and a whole array of govern-ment programme funding, are not part of National Statistics. If significant amounts of public funding are allocated on the basis of the rankings produced by these indices, then, for the sake of transparency and accountability, it is fair to question the quality of the data included in these indices and the reasons why they fail to gain the seal of National Statistics. The differentiation between National Statistics and other official statistics potentially obscures public vigi-lance in relation to data quality. Other than the professionals and those heavily involved in statistics, very few policy-makers will notice whether a dataset avail-able on the ONS website is quality proof or not. They have to look out for the notes in the meta-data with regard to the health warnings attached to the dataset.

Eight months after the publication of the White Paper, the *Framework for National Statistics* (HM Treasury 2000a), together with the *Initial Scope of National Statistics* (HM Treasury, 2000b), were drawn up by the Treasury to set down the arrangements of the National Statistics framework. Whilst the establishment of the Statistics Commission is widely regarded as a landmark development, the Chancellor holds the power of appointing the members of the Commission as well as being responsible for National Statistics. There are, nonetheless, some encouraging signs as the list of statistics initially defined under the scope of National Statistics is wide ranging. More importantly, in some cases all statistics published under the GSS logo have gained wholesale coverage under the new National Statistics framework. The practice does, however, vary widely from department to department.

The irony of this overhaul is that it initially lifted expectations from all fronts, but ended up with shattered dreams. The final outcome of the changes introduced could be dismissed as minimum, and once again the infrastructure of official statistics is heavily dependent on the professionalism and integrity of those who are involved. The reluctance to introduce more rigorous reform and statutory changes, according to the government, is attributed to the length of time and resources involved. However, cynical commentators may think otherwise, especially in the light of the silent rush to publish the White Paper. Even if this official reason is accepted, it is clear that the technical dimension of indicator research is tightly constrained by the wider political and institutional process of data management and organisation at the highest level. Things have, however, moved on since December 2005 after the Chancellor made an announcement to the Confederation of British Industry annual conference that he would publish plans in early 2006 to make the ONS independent. Such a U-turn has been widely welcomed by the RSS and the wider research and policy community.

NEIGHBOURHOOD STATISTICS UNDER ONE ROOF

Since the arrival of the Labour government in 1997, the evidence-based governance ethos has exerted strong pressure on local policy-makers to make an effort to collect and analyse information on a systematic and regular basis. At the same time, the difficulties and unnecessary duplications of data collection are widely acknowledged in the Social Exclusion Unit's *Better Information* report (SEU 2000). There is now a general understanding in government departments that they need to avoid putting the burden of extra data collection on other organisations and that there is a need to develop co-ordination and co-operation over data sharing and data collection (see Wong *et al.* 2004). Following the recommendations of the *Better Information* report, the ONS and other government departments have undertaken significant development work on neighbourhood and small area-based statistics.

To start with, key statistics are much more organised and centrally located in the neighbourhood statistics website of the ONS. The website currently organises statistics by both key topics and geographical areas. The topics covered include the 2001 Census, access to services, community well-being and social environment, crime and safety, economic deprivation, education, skills and training, health and care, housing, indices of deprivation and classifications, people and society, and work deprivation. Each dataset can be viewed online or downloaded in various formats. The datasets also cover some survey-based data such as the New Earnings Survey at the local authority district level, although with a lot of health warnings in relation to the quality of data, largely due to the small sample size involved. There is meta-data for each dataset to describe the purpose and scope of the data, the background and method of data collection, and the caution required in using the data. Another useful service offered by the neighbourhood statistics website is the interactive mapping function. This allows users to obtain statistics for a selected area by simply entering a postcode or a place name. After choosing an area, the location will be visually shown by the automatic zooming of the map on screen.

One interesting observation is that, other than Population Census data and a few other datasets, most data series on the ONS neighbourhood statistics website are published at local authority district level. This poses the question of whether data at local authority level should be counted as small-area statistics and how far the neighbourhood statistics initiative will take us further into the goal of having more fine-grained information to inform urban regeneration and community planning. These questions remain to be answered, but it is somewhat too early to judge the initiative. As the ONS admits, the development of data infrastructure at small-area level is a challenging endeavour which requires tremendous time and effort. This is best explained in one of the newsletters for neighbourhood statistics,

> The NeSS [Neighbourhood Statistics Service] project is making as much data as possible available from a variety of sources, including government departments, local authorities and emergency services. Some data suppliers use computer systems and software which won't work with those used by the Office for National Statistics, where data are made ready for release to the web. In addition, many different ways to categorise data are used, such as by region, county and local authority.
>
> (ONS 2002: 5)

Another key observation is that the Neighbourhood Statistics initiative is very much driven by the *Better Information* report and its framework of development is closely related to the government's Neighbourhood Renewal Strategy. This signifies a major breakthrough in developing cross-departmental data-sharing practice in the government and in underpinning a massive data collation exercise with policy needs. The Director of Social Statistics at the ONS, Karen Dunnell (2002), went into great length to explain concepts and frameworks of neighbourhood statistics in a paper, and concluded that:

> The fact that the statistical system is being developed in such close conjunction with Neighbourhood Renewal policy makes the framework important. It acts to balance

the pragmatic tendency to populate the system with statistics just because they exist. But it also prompts the search for 'good enough' statistics in domains or parts of the framework where there are gaps. It will also inform the longer-term development of neighbourhood or small area statistics. The close link with policy will also stimulate development in new areas and may lead to the identification of new 'domains' over time. The framework also emphasizes the importance of quality, consistency, and the measurement of place and time.

(Dunnell 2002: 17)

The co-location of neighbourhood statistics under one website provides a head start in delivering a more efficient information access service. As the Neighbourhood Statistics project very much addresses policy concerns over social exclusion and social justice, there are many other official statistics that are not covered by the initiative. This means that many useful data sources are still held by individual government departments. With the advent of Internet technology and associated browsers, government departments tend to publish their routinely collected statistics on their respective websites. The navigation of these websites is, however, not all that user friendly. In some cases, the search requires some expert knowledge in knowing where the statistics are located as the data may be hosted in a particular division of the department. As expected, the details and approaches to the documentation and storage of these data sources are not at all consistent.

The one very positive message that comes out from the Neighbourhood Statistics initiative is that at last technical statisticians and policy-makers are talking to each other. It is this interplay of technical and policy needs that pushes the agenda of Neighbourhood Statistics development. However, there is still a long way to go before more small-area datasets are made available to serve other policy purposes. This echoes the discussion in Chapter 2 that researchers have to play an important role in lobbying and advising politicians on the importance of building up the data infrastructure to produce more valid and robust indicators to serve different policy needs.

EMERGING REGIONAL INTELLIGENCE

Apart from the major changes undergone by both National Statistics and Neighbourhood Statistics, there is also an emerging trend of adopting a partnership approach to deliver regional intelligence. After the Labour government was in power in 1997, the pursuit of regional planning agendas was in full swing and consecutive national policy guidance documents have been published, outlining new policy frameworks for economic development (DETR 1997c, 1999a) and physical planning (DETR 1998c, 1998d) in the English regions. The need for statistical indicators to inform regional development was consistently mentioned in these documents. As discussed earlier in Chapter 2, although interest in using indicators to inform policy decisions has been apparent in the urban regeneration and environmental management fields since the late 1980s, there has been a lack of attention on the policy context and organisational approaches to the utilisation of indicators. Such a gap in this area of research raises concerns over the readiness of regional actors to develop the capacity to handle regional intelligence and the extent to which the information will be used in an effective manner to inform policy decisions.

The idea of developing regional intelligence capacity was mooted in the 1997 *Building Partnerships for Prosperity* White Paper. The White Paper stated that 'Regional Development Agencies will build up an expertise in analysis of the regional economy, both theoretical and practical, informed by their extensive contacts with all regional partners and regionally significant sectors and firms' (DETR 1997c: 28). Although the White Paper suggested that a Regional Development Agency (RDA) had to develop regional intelligence capacity, it did not offer any concrete guidance on how to do it. This has created uncertainty on how to accomplish the task. The pressure of developing some kind of regional intelligence organisation has been mounting, as central government places strong emphasis on evidence-based policy-making and imposes monitoring targets over regional strategies. The need to grasp the complexity of the task led to the setting up of a first wave of regional observatories (ROs) headed by the East Midlands in 1999 (EMRA 2001). Others have followed suit, though some are operating in the form of a network rather than a centrally co-ordinated approach. Examples of the network approach include the South West Regional Observatory, which opts for an organic modular approach of building up the policy intelligence base from different partners. The more centralised, co-ordinating approach includes the North West Regional Intelligence Unit, which has the remit of co-ordinating the region's intelligence as well as commissioning new research and data collection (Baker and Wong 2001).

Meanwhile, the need for greater coherence in regional data management was also acknowledged by the government through the creation, in 2000, of the RDA Information Branch at the Department of the Environment, Transport and the Regions (DETR). There has been recognition that regional data are relatively inaccessible in comparison with national statistics. This is partly related to the fact that regional statistics are collected by different organisations at national and regional levels, and there is a lack of co-ordination over the dissemination and publication of the multiplicity of data sources. One response to this problem was the commissioning of a regional data catalogue project to map the availability and use of regional intelligence data by the

Local and Regional Government Research Unit at the DETR. The Institute for Employment Studies at the University of Warwick was commissioned to carry out a comprehensive review of data availability, to identify data gaps (see Green and Owen 2002) and to develop a data catalogue (see Green *et al.* 2002) to provide guidance on effective use of regional and sub-regional data.

Besides the top-down initiatives, major development has been made at the regions. In spite of the needs to comply with the norms and values of evidence-based policy-making, there are variabilities in the observatories' structures and functions across different regions. Although the ROs are led by partnerships involving RDAs, Government Offices and Regional Assemblies plus other bodies, the lead partner largely rests with the RDA, while only a few are led by the Government Office or the local authority-led Regional Assembly. The structure of ROs is largely a function of the interaction among the key partners who have a vested interest over the development of regional intelligence capacity. The dominance of the RDAs over the development of the ROs was found largely due to the resources they provided (Baker and Wong 2004). However, there are emerging signs to show that the ROs are themselves starting to align their work activities by taking into account the enhanced monitoring requirements of the recent planning reforms, positioning themselves in respect to a wider audience of regional partners and stakeholders, and gaining collective recognition via the establishment of the Association of Regional Observatories (ARO) in December 2002. The ARO is charged with the mission to promote 'the best data and intelligence for England's Regions' (ARO website 2003) and serves three main functions:

- To jointly promote and encourage access to the work of the regional observatories.
- To encourage joint working of regional observatories where this will produce added value or lead to savings.
- To identify and promote good practice in provision of regional data and intelligence.

In the paper presented at the Annual European Schools of Planning Congress, Baker and Wong (2004) provided a detailed account on the recent development of regional intelligence capacities in English regions in terms of their organisational structures and potential role in enhancing strategy integration. Some of the key findings in relation to the development of regional intelligence are discussed here.

One of their key findings is that following a strong push by central government towards more evidence-based policy-development at all spatial scales, the organisation and management of regional intelligence and monitoring systems has thus become an important element of evolving regional governance. These rapid changes have, however, created a volatile, uncertain environment within which the recently created ROs have been struggling to meet the demands of their sponsors and partners. This leads to two major trends of development. First of all, regions with a stronger commitment towards regional intelligence capacity building (e.g. North West and West Midlands) tend to adopt a more formal organisational structure or of their ROs and have dedicated staff teams to manage the activities of the observatory. In addition, ROs have needed to reflect on their aims and objectives, establish priorities and continually adjust their organisational structures. This could be clearly seen in terms of the changing involvement of the ROs over their involvement in the preparation of the Regional Planning Guidance, and even greater change might be expected in response to the government's latest reforms to the planning system at the regional and sub-regional levels.

In common with other aspects of the government's regional reforms, the apparent devolution of powers and responsibilities from the centre to the regions is, however, not quite as straightforward as might at first be supposed. As well as setting the overall remit of the RDAs and other key regional players, central government still has ultimate hold of much of the purse strings and, through legislation, regulations, policy and guidance, controls much of the context within which the regional players operate. In addition, the Government Offices, acting in many respects mainly as the regional arms of central departments, are themselves strongly linked to regional strategy development and the development of associated monitoring and intelligence structures, being instrumental in the establishment of ROs in some regions and being represented at ARO meetings (ARO 2004a). All these suggest that the organisation and management of regional intelligence is under strong stewardship of central government.

Such centralised driving forces are, however, met with the undercurrent of a regional observatories movement. In reacting to the instability of this external environment, the ROs have themselves begun to interact with each other, sharing and responding to the problems of uncertainty, and trying to create a common sense of purpose and identity. This is most clearly seen in the establishment of the ARO and its preparation of a Business Plan (ARO 2004b) as part of a drive to lobby central government in terms of both resource allocation and towards improved data collection and analysis at the regional level. They have also recently published *The State of Regional Research* report (ARO 2004c), which is a compilation of extracts from articles and research reports to provide an overview of relevant regional intelligence. Perhaps somewhat unexpectedly, there are thus signs of the ROs becoming important regional institutions in their own right, trying to establish their own agendas and influencing regional partners, national organisations and government in the process. The interaction of top-down and bottom-up driving forces will no doubt continue to shape the development of ROs, which may not be what the original White Paper envisaged.

The development of ROs, as explained above, epitomises the evolving governance structure in English regions, which is a product of continuous power struggle in the central and regional political arena. One aspect of this can be seen in the growing involvement of the ROs in the monitoring frameworks for various regional strategy exercises. Although, as main sponsor, the RDAs can generally be seen to have a primary role in establishing the remit and organisational structures of the ROs, there is some interesting evidence emerging in at least some regions that some of the ROs are also becoming more closely linked with Regional Assemblies and other regional partners, for example through their involvement in the identification and measurement of regional contextual indicators and their role in establishing overarching regional strategies based around Regional Sustainability Development Frameworks or other forms of integrative regional strategies.

Finally, despite evidence of progress in the development of regional intelligence in most regions, there is still great variability in the resources available, the sophistication of monitoring arrangements and the degree of integration with strategy formulation. While the idea of having better co-ordinated regional intelligence is a step forward towards a more integrated approach of governance, there is a need to further examine whether regions are fully committed to such partnership arrangements or whether they are primarily using the observatory idea as a rhetorical gesture to meet central government guidelines. Such partnerships could, however, provide a means of greater standardisation across related regional strategies within a region. If the ARO network is working, we could even expect to see a certain degree of inter-regional harmonisation of data collection practice.

POLICY, POLITICS AND STATISTICS PROVISION: THE ALLSOPP REVIEW

The changing policy agenda of the government at different levels has been constantly shifting the boundaries of the way public datasets are collected and managed. The direction and priority of data infrastructure development are no doubt increasingly subject to the politicisation of competing policy needs and biddings. This could potentially lead to an asymmetrical landscape of data development. With limited resources available and growing policy demands for different types of data, it is inevitable that there will be a need for greater co-ordination and harmonisation over data sharing and collection at all spatial levels in the near future.

Recently, the Chancellor of the Exchequer, the Governor of the Bank of England and the National Statistician commissioned a major review (Allsopp

2004) of the information requirements for monetary and wider economic policy-making, with a more specific remit to assess the demand for and provision of regional information. The Allsopp Review, which is supposedly to be a technical review, offers a glimpse of the interaction between politics, policy and the data provision agenda. Throughout the report, it is obvious that different political forces are in play, which exert pressure for particular policy needs that are not easily met by the existing statistical infrastructure at the national and regional level. The pressure mainly comes from the European Union and the devolved regional administration structure. This is clearly reflected from the term of reference for the Review:

> to advise on changes in the statistics and information necessary if the UK were to join the European single currency ... the regional information and statistical framework needed to support the Government's key objective of promoting economic growth in all regions and reducing the persistent gap in growth rates between the regions; and whether the changing economic structure of the UK is being properly reflected in the nature, frequency and timeliness of official economic statistics.
>
> (Allsopp 2004: 21)

The European Commission's requirement from Member States largely comes from its need to assess their contributions towards the EU budget and to monitor their performance against agreed policy initiatives such as the qualification for Structural Funds (Allsopp 2004: 49). In 2000, in Lisbon, Europe's leaders further committed themselves to a 10-year strategy of reforming Europe's labour, capital and product markets. This means that any new or revised EU legislation will impose pressure on the resources of the statistical services and may add the risk of distorting the priorities and pre-empting the resources ear-marked for other statistical services. The contentious relationship brought by supranational legislative requirements and national statistical strategies is clearly shown in the report. Hence, Allsopp and his team made Recommendation 73 that:

> any additional financial and compliance costs of new statistical regulations introduced by Eurostat should be transferred from the budget of the current government department that leads on the relevant Council formation to the ONS, once the ONS has taken on the measurement role.
>
> (Allsopp 2004: 154)

The prominence of European influence is also extended to the standardised geographies used in compiling statistics. The Nomenclature of Units for Territorial Statistics (NUTS) geographies of Eurostat were accepted by the Review as the standard and the estimates of regional economic activity at levels below (regional) NUTS 1 areas have added importance because they are the basis for decisions on allocating the EU Structural Funds.

The tension between competing policy needs and the resources available to meet with such needs is also found at the national level. While the Allsopp Review placed the development of better-quality regional Gross Value Added (GVA) estimates at the high priority, the ONS's response failed to satisfy the Review Team. Although recognising the knock-on impact on the ongoing modernisation programme in the ONS, the Review made it emphatically clear about the urgency of reforming the provision of regional data:

> The ONS' response stressed the need to view development of Regional Accounts as part of the ongoing modernisation programme. However, its timetable suggested that the development of sources and new methodology would take several years, with experimental data on the new data for 2006 becoming available in 2007–08, alongside existing data.
>
> (Allsopp 2004: 141)

> The Review Team is not in a position to assess the competing priorities for ONS resources or to speculate on funding levels. Nevertheless, the proposed timetable is likely to disappoint many users of regional data. We would challenge the ONS to look for areas where earlier progress might be possible, including in particular the development of the timely indicator of regional GVA.
>
> (Allsopp 2004: 142)

There is also a sense of uneasiness over the contest of different types of statistics over ONS's resources. As discussed earlier in this chapter, Neighbourhood Statistics were developed to meet with the government's agenda in tackling social exclusion. The information has been widely used in neighbourhood renewal initiatives under the Office of the Deputy Prime Minister. The implication from the Allsopp Review is that there is a need to rebalance the ONS resources between social and economic statistics. As the remit of the Allsopp Review is on economic policy-making, it is thus not surprising that it fights for economic statistics' corner in the report:

> Recommendation 22: The considerable work undertaken in recent years to develop the Neighbourhood Statistics Service shows what can be possible: although not without a price. We support the aim of the ONS to continue development of the Service. We recommend that this, and parallel systems in devolved administrations, should include scope to cover an expanded range of economic data, which could be presented at a range of different geographies below and up to NUTS 1 regions, to become the key central resource for micro-regional data. . . . The range of different data, including regional economics and local neighbourhood, might suggest a suite of different

access portals for such an expanded system. But the 'Neighbourhood Statistics'
badge should be retained for those data most relevant for neighbourhood renewal.

(Allsopp 2004: 149)

While the Allsopp Review provides some helpful directions to the development
of regional and national statistics, its strong economic bias however does not
help to provide an all-rounded view of how statistics can contribute to the overall
welfare of the society. These limitations are recognised by the Review Team that
the welfare agenda is 'better judged, not in terms of economic growth, but in
terms of quality of life.' (Allsopp 2004: 54). Such a compartmentalised review
also fails to provide a balanced view of the synergy of different strands of statistics towards joined-up policy development. As admitted in the report: 'Some of
the [social and environmental] data will help inform economic policy making,
while economic variables will inform social policy. However, these indicators and
others, including, for example, environmental indicators, are largely beyond the
scope of the Review' (Allsopp 2004: 54). Such a fragmented approach of statistics review also makes it difficult for the ONS to prioritise its policy needs, and
its limited resources will continue to be a political football ground where different
levels of legislative requirements from different government departments have to
contest. It is, nevertheless, clear that with the ONS directly responsible to the
Chancellor of the Exchequer rather than to parliament, the priorities of resources
could be distorted. This is exactly the concern expressed by the Royal Statistical
Society over the half-way reform proposed in the *Building Trust in Statistics*
White Paper.

PART II

CONCEPTUAL, METHODOLOGICAL AND ANALYTICAL ISSUES

CHAPTER 5

DATA: A REQUIREMENT AND A PROBLEM

DATA CHALLENGES

Of all the stumbling blocks in indicator research, it is clear that it is 'data, data and data' which makes it or breaks it. Data is both a requirement and a problem to indicator development. Without the basic ingredient of good-quality datasets, it is simply not possible to produce reliable and robust indicators, though in some cases innovative methodology and analytical techniques can help to ameliorate and overcome some of the problems. The rapid development of information technologies, notably the application of geographical information systems, the affordability of personal computers and the development of spreadsheets, database managers and various user-friendly statistical packages has largely enhanced our information-handling capacity in the last twenty years. The real concern is, however, how to capture reliable and good-quality information efficiently and effectively to provide the basic ingredient for analysis. Following the recommendations of the *Better Information* Report (SEU 2000), significant development work on neighbourhood and small-area based statistics has been carried out by the ONS and other government departments. To start with, key statistics are now much more organised and centrally located in the neighbourhood statistics website of the ONS. With the advent of Internet technology and the World Wide Web browser, many government departments also publish their routinely collected statistics on their websites. While this progress in improving statistics is encouraging, there are still plenty of challenges ahead. This chapter, therefore, devotes attention to examine issues surrounding data availability and data quality, and to identify the inherent problems in current public data compilation practice. It also explores how methodological research interacts with the policy agenda to overcome some of these challenges. The discussion is augmented by the experience gained from compiling data for two indicators projects: local economic development indicators study (see Wong 2002a) and town and city indicators database research (see Wong *et al.* 2004).

UNCERTAINTY AND VOLATILITY IN THE DATA FIELD

Whilst the progress in improving statistics is encouraging, it does introduce a sense of uncertainty and volatility to the data field. The experience of compiling indicators for the Town and City Indicators Database (TCID) over a 9-month period has been a steep learning curve to the project team, as the methods of collecting and compiling various national statistics have been under constant revisions and adjustments. During the time period in which the research was conducted, there were revisions to the published Annual Business Inquiry (ABI) employment data. This meant that some employment-related analyses conducted at an earlier stage of the project were dropped in favour of analyses incorporating the new information. Similarly, during the time period of the project, mid-year population estimates for 2001 were released, and subsequently revised, as were mid-year estimates relating to earlier years. Meanwhile, local estimates from the Labour Force Survey were also re-based in accordance with results from the 2001 Population Census. In January 2003, further changes to the definition of local unemployment statistics were introduced. All of these changes symbolise the dynamism of the information base. The construction of indicators has increasingly turned into a game of jigsaw that requires patience and knowledge to put the pieces together. With all the changes involved, the interpretation of some indicators, especially change measures, has to be cautious. Such changes make it unreliable to identify any emerging trends or to conduct comparative analysis. This reinforces the message that the indicator value is 'indicative' rather than exact science.

From our recent experience, while it is clear that each data source has its advantages and drawbacks, the choice of a particular dataset very much depends on the characteristics of the case in hand. The problem of constant revision of methods for collecting and reporting key statistics, such as employment and unemployment, is profound and well documented. For instance, many changes have been made to the methods used to compile and report unemployment statistics since the 1980s, which have already caused serious concerns over the independence of statistics from political interference (Bartholomew 1995; Wong 2000). The revisions made to the employment data series have been notorious (Wong 2003): both in terms of their collection methods and the definitions used. In the last decade, the collection of employment data has shifted from a bi-annual census to an annual survey, and, subsequently, the employment surveys were replaced by the business surveys (ABI). This partly reflects that the priority of the government is to spend resources to collect more updated information, though at the expense of spatial accuracy especially when examining employment structure at the local level. It is thus very interesting to note that in the recent review of statistics for national economic policy, the Allsopp Report was very critical of the changes and inconsistencies found in the methods of estimating labour market statistics. The report commented that:

The main problem is that, in practice, there are significant differences between the number of jobs estimated from household surveys (the Labour Force Survey), and from business surveys (the ABI and intermediate Workforce Jobs updates of this). These differences are both in terms of levels and changes over time.

(Allsopp 2004: 116)

While it is possible to argue that each of the separate sources of data has some advantages in its own right, more needs to be done by the Office for National Statistics to explain the reasons for differences between the Labour Force Survey and business survey sources. This was the motivation of the Quality Review of Employment and Jobs, begun in June 2003.

(Allsopp 2004: 118)

Since employment and unemployment data are the two most widely used information sources to ascertain socio-economic change, any revision made to these data series can have serious implications for policy monitoring. This is because the wider impacts brought by some policy outcomes may require ten to twenty years to reveal. In spite of the effort made by the Office for National Statistics (ONS) to revise some past data series, such revision tends to go back only a few years, which is not far back enough to allow a sufficiently long period of time to make reliable observations. For long-term trend analysis, the only way forward is for researchers to amalgamate different statistical series together (e.g. Begg *et al.* 2002) and try to make sense out of the lot. Without a consistent basis, researchers and policy-makers will be unable to attribute the extent of change that is genuine from that caused by a switch of methodology. It is hard enough to isolate the additional effect brought by any policy initiatives to develop performance measures, let alone deal with the extra noise created by the inconsistency of statistical series. The changes made to the unemployment and employment data series also raise the issue of trust and confidence over public data compilation practice. Local policy-makers have expressed doubts and concerns about the reliability of key data sources and sometimes prefer to use their own information sources. It is, however, not feasible to make use of bespoke local sources to carry out nationwide comparative analysis. If statistics are to be used to provide useful policy intelligence, the statistics castle has to be built upon a solid rock rather than a shifting sand dune. This concern was echoed in the recommendation made by Allsopp:

We consider the prospect of a single series of jobs data to be a reasonable aim and recommend that the ONS reviews the work required to develop a single measure, in the light of implementing the proposals from the Employment and Jobs Quality Review.

(Allsopp 2004: 120)

DATA ACCESSIBILITY AND DEVELOPMENT ISSUES

As discussed in Chapter 4, National Statistics and other official data remain the most important information sources when compiling indicators. The advantage of using these official datasets is their credibility as relevant data, which are thus more readily accepted by the users. During the process of compiling the Town and City Indicators Database, some administrative databases designed for specific policy purposes were found to contain information that could be processed for reporting at finer spatial scales rather than just for local authority districts. However, in many cases, the release of such data at the detailed spatial scale for the measurement of urban areas would require a significant amount of research effort to process, refine, validate and develop the data to ensure reliability and consistency. Due to confidentiality or other reasons, these datasets tend to have restricted access. This means that special arrangements may have to be made to negotiate with the relevant data holders to set up research programmes to further refine and develop these datasets in order to maximise their potential for application to monitor urban performance and progress. Such datasets include planning application statistics, land use change statistics and air quality data.

The 'secondary' use of administrative data is becoming increasingly widespread, even though these datasets may not carry the 'quality proofing' of national statistics. The main reason for the growth of this kind of information has been the burgeoning number of indicators for monitoring public service delivery, such as the Best Value performance indicators of local authorities. As the potential value of administrative sources is increasingly recognised, there is scope for improved information on a number of topics. For example, detailed information on receipt of a range of benefits is already obtainable. There is the prospect of enhanced information on educational achievement, following the successful implementation of the Pupil Level Annual Schools Census and development of a National Pupil Database. In addition, routine administrative records such as the Council Tax database held by local authorities will offer tremendous potential for up-to-date spatial information on the issue of housing voids and vacancy. In order to develop the use value of these datasets, a significant amount of resources and political will may be required to process the data so that it can be released in aggregated spatial forms without breaking the rules laid down in the Data Protection Act (DPA). The frustration is that many data sources required to measure policy issues are actually there, but they are not processed or disseminated in a form that could be accessible for others to use. This is partly related to the misunderstanding of the nature of the DPA. It is no doubt that the DPA is a highly technical and legally bound area, and, on occasions, a strict blanket interpretation could rule out activities that would otherwise be acceptable (SDRC 2004). As the PAT18 report very helpfully pointed out, the publication of aggregated statistical information from which individual information cannot be deduced is not blocked by the DPA (SEU 2000: 17).

Data accessibility still remains as an issue when the data is not covered by any information from a government source or, perhaps more frequently, the datasets are only covered by government surveys that do not provide a more fine-grained spatial breakdown. With the pressing need of having more updated information on various indicator sets, the use of survey data to construct indicators becomes a major topic of discussion. The trend of 'modelling' data to estimate indicator values to local authority wards has been started in the Index of Multiple Deprivation (IMD 2000) (DETR 2000b). Three different types of modelling were used in producing ward-level estimates of data:

1   Survey-derived estimates: national surveys such as the English House Condition Survey and the Labour Force Survey were used to derive indicator values for wards. Examples include the poor private sector housing indicator and the working-age adults with no qualifications measure.
2   Attributing estimates: this normally involves the use of some criterion variables to directly attribute district level data to wards with the total figure controlled at the district level. Examples include the estimates of pensioner and disabled recipients of Council Tax Benefit in the IMD 2000.
3   Allocated values: the indicator value of wards will be exactly the same as that for the entire district; this was applied to the Comparative Mortality Ratio indicator in the IMD.

For survey-based estimation methods, geodemographic area classification schemes tend to play a major part in the modelling process. The ONS currently holds a licence of the 'A Classification of Residential Neighbourhoods' (ACORN) scheme, which is used as the basis to develop the sampling frame of many major government surveys including the British Crime Survey and the English House Condition Survey. The ACORN classification is largely based on cluster analysis of the Census data and other information. For example, the 1991 ACORN classification consists of thirty-eight neighbourhood types and eleven neighbourhood groups (see Table 5.1). Since ACORN was used to develop the sampling strategy of various public surveys, it is thus not surprising to see survey findings presented as aggregated data for ACORN groups. For example, Forrest and Kearns's (2001) analysis of social cohesion was widely drawn upon survey findings that linked to ACORN area types. These area groups are, however, too crude to offer much value to develop spatial indicators for small-area analysis. Facing these difficulties, one approach adopted by researchers is to model survey data values from these broad area types into smaller spatial units. One example to illustrate this modelling approach is the methodology developed by the University of Liverpool to estimate crime data for local authority wards. The research team used the Home Office's Basic Command Unit (BCU) crime statistics to estimate crime counts and rates at the ward level for the Association of London Government (Robson *et al.* 2002). The estimation procedure used the distribution of the residential population of wards across different ACORN categories to estimate the likely share of the BCUs crime that would fall within each ward. This was done separately for property and personal crimes, using information from the British Crime Survey on the relative risks of these types of crime and on the relative recording rates, for the different ACORN categories.

There are several concerns over using such an approach to derive indicator values. Regardless of which geodemographic classification scheme is involved, there are reservations over the use of the technique of cluster analysis in carrying out area classification. The technique is subject to the judgement of the analyst in terms of the choice of data, clustering method, number of clusters used and labelling of the cluster (e.g. Openshaw *et al.* 1985; Webber and Craig 1976). This raises the issue of transparency in using these commercial products, as there is a statistical black box over the methodology used and the choice of different variables and data sources. The methodology and the details will not be available in the public domain for scrutiny and debate. The use of survey data based on geodemographic systems is also not straightforward and makes comparison difficult when different classification schemes are used. It is also important to note that geodemographic products are developed for marketing purposes and have a strong commercial interest in mind, which may not be that appropriate to be

Table 5.1 Neighbourhood groups in the 1991 ACORN classification

A	Agricultural areas (3.3% of 1991 GB population)
B	Modern family housing, higher incomes (16.8%)
C	Older housing of intermediate status (17.8%)
D	Older terraced housing (4.2%)
E	Council estates category I (12.7%)
F	Council estates category II (9.0%)
G	Council estates category III (6.1%)
H	Mixed inner metropolitan areas (3.4%)
I	High status non-family areas (4.7%)
J	Affluent suburban housing (17.7%)
K	Better-off retirement areas (4.2%)

Source: CACI (www.caci.co.uk/acorn/).

used for the sample design of deprivation and other public policy-related issues. Furthermore, the adoption of any product developed by commercial companies will inevitably involve significant data purchase costs, especially when the product is subject to revision and updating.

While the use of survey data provides valuable opportunities to broaden the data sources available for measuring indicators, it is however important to recognise that the use of survey data requires considerable effort to establish the robustness and consistency of the modelled values between areas and over time. The recently released 2001 Population Census data offers an opportunity to validate some of the sample survey findings with that of the Census Output Area (the smallest census reporting unit) data. In a recent review of updating the deprivation index for Wales, Robson *et al.* (2003) assessed the validity of the geography of results from sample surveys with the small-area data from the 2001 Census. They compared the results of the Local Labour Force Survey (LLFS) with those of the Census, both in 2001. The test indicator chosen is 'lack of qualification', though the definition used in the LLFS is somewhat different from that in the Census. The LLFS variable measures the percentage of people of working age with no qualifications at 16–64 for males and 16–59 for females. In the Census, the variable is specified as people aged 16–74 with no qualifications. It is, therefore, not surprising to find that the LLFS percentage is 21.2 for Wales, which is considerably lower than the Census data at 33.0. The percentages for the twenty-two unitary authorities in Wales are shown in Table 5.2. In spite of the differences in overall levels, there is nevertheless a close relationship between the Census and LLFS data for authorities. The closeness of fit would give one some confidence in using the Census small-area data as a way of allocating the unitary authority-level values from the LLFS to more fine-grained spatial scales. It is, however, important to recognise the strengths and weaknesses of such modelled values in policy terms. While modelled indicator values can serve the purpose of deriving values to facilitate the measure of *relative* needs in the funding allocation formula, they are not that robust and reliable to allow longitudinal analysis of trends and changes.

FINE-GRAINED AND UP-TO-DATE SPATIAL DATA

Most indicators used to inform urban and regional development tend to be geographically based; the choice of a spatial scale appropriate to the problem is thus very critical (Archibugi 1998). Some issues such as environmental improvement are best dealt with at the neighbourhood level, while others such as infrastructural capacity are more appropriately measured at city or regional scales. The choice of scale of measurement is, however, constrained by the availability

Table 5.2 Percentage of working-age people with no qualification, 2001

Unitary authorities	% working-age people with no qualification – Census	% working-age people with no qualification – LLFS
Anglesey	31.8	19.4
Gwynedd	30.0	19.9
Conwy	31.8	17.7
Denbighshire	31.3	19.0
Flintshire	29.3	19.7
Wrexham	33.2	24.7
Powys	31.3	21.3
Ceredigion	24.9	19.0
Pembrokeshire	31.1	19.2
Carmarthenshire	34.0	22.8
Swansea	30.5	16.9
Neath Port Talbot	39.0	29.6
Bridgend	36.4	23.5
The Vale of Glamorgan	26.1	15.2
Rhondda Cynon Taff	40.5	22.7
Merthyr Tydfil	43.9	27.3
Caerphilly	39.7	28.0
Blaenau Gwent	45.0	33.9
Torfaen	36.6	25.4
Monmouthshire	26.3	14.2
Newport	33.5	22.1
Cardiff	26.8	16.7
Wales	33.0	21.2

Source: Robson et al. 2003: 44.

of existing statistical data. There is a trade-off between the amount of data available and the use of more appropriately defined spatial units. Administrative boundaries are often used as a framework for data compilation, but they may not correspond to the ideal spatial scale of measurement for a particular phenomenon (Carley 1981). Indicator values are highly sensitive to the definition of the spatial units for which data is aggregated. By altering the definition of the boundary, the outcome of the analysis may also change. Openshaw (1984) called this the modifiable areal unit problem. Figure 5.1 illustrates that by altering the boundaries of the wards in a local authority, the deprivation indicator value of each ward can vary dramatically.

Recent regeneration policy has shifted to place a strong emphasis on the bottom-up, community-led approach as evident in the advent of the Neighbourhood

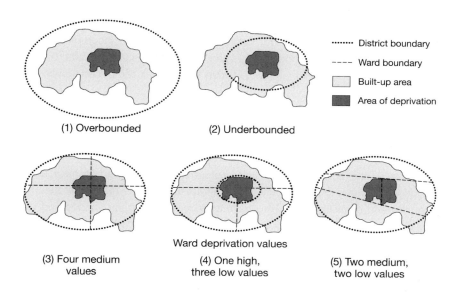

Figure 5.1 Modifiable areal unit problem

Source: Robson 2003.

Renewal Fund and the Local Strategic Partnerships. The provision of spatially disaggregated data to meet with policy needs has thus been regarded as one of the most critical issues in affecting the development of policy indicators in Britain (Anderson 1998; Wong 2002a). In compiling the Town and City Indicators Database to examine urban change, only thirty-nine out of the ninety recommended indicators are available at fine-grained spatial levels (i.e. wards and postcode sectors) to allow the development of indicators to cover all 257 Primary Urban Areas in England. The remaining fifty-one indicators, by and large, were calculated using datasets available at the unitary authority/local authority district level and can only cover seventy-eight of the 257 Primary Urban Areas. Many of the useful data to measure urban change are administrative data, such as the Housing Investment Programme returns, Land Use Change Statistics, Planning Application Statistics, Valuation Office Agency Statistics, Best Value indicators, school examination results and the National Land Use Database (NLUD). However, with the exception of the NLUD and school examination results, the other data sources are released at the district level. There is a strong prospect to further develop the Land Use Change Statistics and the Valuation Office Agency Statistics to allow the release of more detailed infor- mation. As discussed above, this will require special arrangements with data holders and separate research programmes to develop the statistics for release at more refined spatial levels. Following the devolution of Scotland and Wales, it is also increasingly difficult to obtain data for Great Britain and the UK as many data sources are only available for England. The problem of incomplete spatial coverage is also affected by the suppression of data (e.g. earnings data for some London boroughs and Welsh authorities) due to the small sample size involved in these areas.

Following the recent publication of the 2001 Population Census, there is now a window of opportunity to obtain small-area data to carry out analysis. One major development in the 2001 Census is that a new set of geography was used as the building block. The lowest data output unit is Output Area. Output Areas are built from clusters of adjacent unit postcodes and designed to have similar population sizes and be as socially homogenous as possible (based on tenure of household and dwelling type). They have approximately regular shapes and tend to be constrained by obvious boundaries such as major roads. With the continuous revision of administrative boundaries such as local authority wards, there is a need to develop indicators from a set of more stable and consistently defined geographies. The Output Area geographies may offer a better prospect of database development in the future.

The Census of Population remains a crucial source of information for spatial distribution analysis as it provides consistent and comprehensive information at the micro area level. Nevertheless, one of the fundamental problems of using Census data is the cyclical obsolescence issue, as the data is available only on a decennial basis. For example, by the time the 2001 Census data is fully released in June 2004, the data is not exactly recent. Further opportunities for better small-area data and Neighbourhood Statistics will then rely on the progress of a programme of work currently being developed by the ONS and the Neighbourhood Renewal Unit. There is thus a mismatch between policy needs and the availability of spatially disaggregated data.

DIFFICULTIES IN MAPPING DIFFERENT GEOGRAPHIES

Even when many data sources are now available on the Internet or in computer readable format, they still require plenty of processing work to convert the data into usable format. Although fine-grained micro data is now available for a whole array of subject areas such as student examination results, train timetables, university research assessment exercise results, motorway accessibility search, derelict and vacant land and house prices, they tend to be published at different spatial geographies such as grid-references, postcodes and Census geographies. In order to develop indicators for policy analysis, these datasets have to be processed and converted into a common spatial framework of analysis via computer software.

The matching between postcodes and Census geographies, for instance, is not always straightforward, and errors have been encountered in the geography look-up files developed by expert researchers. More importantly, suitable weighting factors may be required to aggregate data from one type of geography to another. For example, when aggregating average house price data from postcode sectors to wards, the house price data has to be weighted by the number of properties involved in each postcode to avoid distortion of the uneven number of houses in different postcode sectors. This means that the data compilation process is very lengthy and that researchers have to possess sufficient technical and conceptual skills to handle these complicated tasks. This no doubt poses hurdles to users who are not experts in manipulating data and leads to a trend of commercialisation of information (Coombes and Wong 1994) whereby specialists from the private sector are employed to perform the task.

The private sector and various agencies have increasingly played an important role in processing and analysing policy-relevant statistics. For example, house price data is now available at postcode address points from the Land Registry. Researchers and policy users can purchase the raw data and output it to different spatial geographies to tailor for their policy and analytical needs. However, for those who do not possess the skills to do the task, they can purchase bespoke data for specified spatial units from the Land Registry, though at a more expensive price to cover the cost of research time involved. Another example is the vehicle stock data that can be purchased from the commercial selling agents of the Driver and Vehicle Licensing Agency. Since the processing of raw data into usable formats usually involves technical skills as well as data purchase costs, some private companies such as *Experian* and *CACI* have specialised in developing tailored data products by processing and analysing raw data and then selling them to clients in government departments, local authorities and other private companies. With the restrictive research capacity in local government (as discussed earlier in Chapter 3) and the widely acknowledged shortage of quantitative research skills, the trend of commercialisation of information is very likely to continue. There will be markets for more refined and value-added data products in the public policy arena.

DIFFICULT TO MEASURE ISSUES

Following better data collection practice and innovative use of different data sources, there are now more data sources available to develop policy relevant measures. More data sources, nonetheless, do not resolve the hurdle in finding appropriate data to measure certain issues. The discussion here aims to highlight these difficult areas of development. First of all, there is a lack of appropriate quantitative data to develop satisfactory and meaningful measures for intangible issues such as community identity and institutional capacity. When developing indicators, there is an implicit assumption that we can quantify and measure a particular issue of concern; however, actual experience tells us that this is not often the case. Some concepts such as institutional capacity, community identity and the aesthetic quality of an area cannot be easily measured, because the nature of these subject matters is more concerned with quality rather than quantity and will require subjective judgement and opinion from the public or the analyst. Researchers such as Hemphill *et al.* (2004a, 2004b) have thus employed an expert opinion-based scoring system to measure such intangible issues. This will, however, require significant time and resource inputs as well as research skills to produce reliable measures.

There are also significant difficulties in finding any reliable and robust information to measure financial resources, training and skills of workforce, community participation and certain aspects of infrastructural resources such as telecommunication network and public utilities. This is partly because such information is either regarded as commercially secret and thus not available to the public (e.g. regional energy charges), or will not provide sufficient/consistent details (e.g. venture capitalist directory and register of voluntary and community groups) for indicator creation. The third difficult area of development involves the measurement of environmental indicators at the local level. Many environment indicators are collected at the local level and will not lead to the development of a consistent nationwide database to allow full coverage of all local areas. Certain environmental information, such as air and water quality, is collected on the basis of the location of the sampling stations, which make it difficult to attribute the value areas beyond the immediate vicinity of the stations. As far as the official Quality of Life Count (previously Sustainability Count) indicators are concerned, they are collected (DETR 1998a) at national and regional levels only, though a good practice guide (DETR 2000a) is issued to suggest what could be collected at local authority district level.

INDICATOR HARMONISATION: A MIXED BLESSING

As discussed in chapters 3 and 4, the evidence-based policy ethos has exerted more and more pressure on national and local policy-makers to make an effort to collect and analyse information on a systematic and regular basis. At the same time, the difficulties and unnecessary duplications of data collection are openly stated in the *Better Information* report. The need for greater coherence in regional data management to meet the increasing demand of regional statistics was also acknowledged by the government through the creation of the Regional

Development Agency Information Branch at the Department of the Environment, Transport and the Regions (DETR) in 2000. It is, therefore, not surprising that one of the premises of the Town and City Indicators Database (TCID) project is that indicators would only be compiled from data already available in the public domain without imposing extra burden on collecting new data. Throughout the TCID project, there has been close liaison between the project team, the Office of the Deputy Prime Minister, the Office for National Statistics and other key government departments and agencies to raise awareness of other relevant indicator sets and policy targets, for example Best Value indicators, Index of Multiple Deprivation 2000 and neighbourhood renewal targets. At the local level, many local authorities and their sub-regional and regional partners have been active in developing indicators to measure and monitor aspects of quality of life and conditions. The development of the TCID has been closely related to the development of the latest policy issues and aims to make close links with relevant indicators developed from other policy initiatives to inform progress of urban development. In the near future, it is anticipated that the TCID will be more closely related to the Public Service Agreement targets, Urban Audit II and the monitoring indicators of the Sustainable Communities. All these signify progress towards collaboration and joint working to streamline the collection and usage of information. However, the question is how far should we push towards the ideas of having a fully integrative data infrastructure?

The ethos of evidence-based policy and a trend towards multi-agency working and data sharing inevitably point to the need to harmonise different indicator sets to avoid duplication and different versions of similar indicators (Wong 2000). There may be scope for different government departments to harmonise the usage of indicators across different policy domains. Since there is already a set of deprivation indicators and a set of economic competitiveness indictors, it may be sensible to just focus on developing environmental indicators and make use of these three sets of indicators to see how they can contribute to the four objectives of sustainable development rather than having a separate set of Quality of Life Count indicators. This integrative approach could help to clarify the interconnection between different factors and to fine-tune the choice of indicators. It can potentially create synergy, stimulate debate among politicians and civil servants across different policy spectra, and avoid the current duplication of effort in examining similar issues in different parts of government. While the proposal of a neat and harmonised set of policy indicators sounds attractive, the reality is that different policies place slightly different emphasis on different issues and will require different spatial units of observation. It is, therefore, not easy to impose a 'one size fits all' strategy as it will only lead to the achievement of the 'lowest common denominator' situation without fully acknowledging the monitoring needs of different policy initiatives.

This means that there is a paradox here: on the one hand, we argue for a more coherent and streamlined structure of public statistics; on the other, there is a need to retain flexibility and to avoid the 'one size fits all' problem. The suggestion put forward here is that harmonisation should come at different levels and in different stages. It should perhaps aim at developing a coherent framework of data collection first, in terms of definition, methodology, and spatial scales and coverage, to provide a platform to streamline the public data infrastructure. Once this foundation has been achieved, further development over the harmonisation of certain indicator sets could then be considered. It is very unlikely that there will be a definitive set of indicators to guide policy-making. Nevertheless, for certain indicator sets that are addressing similar issues of concern, harmonisation may be a desirable approach to avoid information overload and potential confusion. There is no straightforward answer to the issue of harmonisation as it is down to the balancing act of maintaining flexibility of policy intelligence and the need to streamline the data infrastructure. More importantly, the process will require strong leadership and commitment to cross the divide between different departments and different policy sectors to achieve integration.

The discussion here shows that data is a key requirement but also a millstone in the process of indicator development. There is obviously tension between the policy needs of data and evidence and what is in store. While good progress has been made over recent years to overcome some of the problems, there is still a long way to go. I would say that we have not even reached the plateau, let alone the summit. At the superficial level, with the launch of the ONS Neighbourhood Statistics website, there is a sense of false optimism towards the prospect of having fine-grained spatial data, especially to the novice of this complicated subject matter. In the light of the discussion made here, there should be optimism, but in a more measured manner. There is also a need to address the issues of reliability and consistency in data collection and to consider the possibility of institutionalising some important statistical sources.

CHAPTER 6

TECHNICAL OR ANALYTICAL SYNTHESIS OF INDICATORS

Indicators alone are idle information, which hardly convey any meaningful message for policy-making. It is the analysis of indicators against the wider context and policy objectives that provides the added value of converting information into intelligence. The focus of this chapter is, therefore, to explore alternative approaches used to improve the interpretation, analysis and presentation of indicators. As discussed in Chapter 2, one of the key concerns of indicator development is how to provide a synopsis of the concept being measured. Paul Lazarsfeld commented that 'so long as a set of data has not been classified, even summarily, it is impossible to analyse the relations between variables' (Lazarsfeld 1970: 329). Olson also stressed the importance of having a consolidated grouping of social indicators by aggregation, by representation or by classification (Olson 1969 in Cazes 1972: 21). The need to provide a parsimonious summary of the meaning of indicators is also a pragmatic one. If research is to be infiltrated into the decision-making process, the message that emerges from the findings has to be sharp and clear. The discussion here aims to explore some of the longstanding debates over the techniques used to simplify indicator values and the pros and cons of creating composite indices to provide the type of policy intelligence required.

WEIGHTING AND COMPOSITE INDICES

The default option used to simplify an indicator set is to combine or aggregate individual indicators into a single composite index, as it provides a hard and fast technical synthesis. This challenge in turn raises the possibility of 'weighting' the indicators according to their relative importance. The weighting scheme used to combine different indicators is indeed very similar to a cooking recipe that specifies the quantity of different ingredients to make a dish. It is always intriguing how the taste of the dish can dramatically change by simply varying the relative proportion of each ingredient used. The logic and consequence of varying cooking ingredients applies when devising a weighting scheme to combine individual indicators. For example, Gordon (1995) derived a set of weightings from the

Breadline Britain Survey through a logistic regression analysis to find out the relative weightings of different variables in estimating the number of deprived households. Saunders (1998) duplicated this methodology for a study in Greenwich, but she found that unemployment carried more weight towards deprivation in the Greenwich sample than in the national sample used by Gordon. The two sets of weighting then produced different estimates of the number of deprived, with a differential of 1,250 people. Such a difference could be important if funding resources were attached to the number. It is thus important to carry out sensitivity analysis (e.g. Coombes *et al.* 1993b, 1995) of the assembled database to identify differences in the outcome produced by alternative weighting approaches before making the final judgement. There are two broad approaches to devise a weighting scheme: non-statistical and statistical methods.

Non-statistical weighting methods (listed in Table 6.1), such as applying unitary weighting, asking expert opinions directly or using an iterative technique (e.g. the 'Delphi method') and deriving weightings from previous literature and public opinion polls, have the advantage of simplicity and are easily understandable. The advantage of simplicity is visibility, which means the decisions on weighting can easily be recognised and debated. However, a simple method is not necessarily a less contentious option because it may not provide the most appropriate answer to policy targeting and is subject to arbitrary and subjective judgement. For example, it is difficult to derive a set of weightings to provide a trade-off between air quality and economic growth. Due to the practical difficulties involved in using the weighting systems mentioned above, an alternative way forward is to focus on a purely empirical assessment of the indicators themselves. Various statistical techniques (see Table 6.2) such as factor analysis, regression modelling and multi-criteria analysis can be used to produce a combined multivariate index from the selected indicators. The downside of these methods is that they tend to be more complicated and create a statistical black box that makes the process less transparent for interpretation.

Composite indices, on the whole, tend to be more appropriate to provide a synoptic overview of issues at a higher spatial scale, but they are less responsive to pinpoint issues at the lower rungs of the spatial hierarchy (Sawicki and Flynn 1996). For instance, Eurostat's (2001a) 'Environmental Pressure Indices' are primarily developed for application at the national level, which are found less useful to inform the progress of sustainable development at local and regional levels. It is thus not surprising to find that sustainability indicator sets, comprising a broad range of indicators, are rapidly emerging in local communities across the world (Innes and Booher 2000). The other concern over the use of composite indices is that they conceal detailed information on different aspects of the phenomenon studied. This is especially problematic when the relationship between the indicator and the phenomenon concerned is ambiguous. For instance, when

Table 6.1 Non-statistical weighting methods

Null: the default method is not applying any weights to the selected measures. The 'Booming Towns' analyses of Green and Champion (1991) provide examples of a preference for applying null weights to the selected indicators. The apparent benefit of simplicity from this approach is also clearly a disadvantage, in that it assumes all indicators are of equal importance regardless of the concept involved, the nature of the data available, or the objectives of any specific policy initiatives for which the ranking is needed.

Expert: another method is to obtain the assessment and opinions of experts in the specific application field. The 'underprivileged area' study by Jarman (1984) is a classic example of using an opinion survey of General Practitioners to devise a weighting scheme to combine eight indicators into a single index to measure General Practice workload. Similarly, the weighting scheme of the Grant Thornton Index was based on a poll of the state manufacturers' associations to measure the state business climate in the USA (Boyle 1989). Expert opinions can be elicited by asking their preferences directly, or using iterative techniques such as the 'Delphi method' where the experts are asked to address a problem anonymously in two or more rounds until consensus is achieved (Sackman 1974). The use of expert weightings has the advantage of integrating practical experience into the analysis. However, it is difficult to decide who are the experts and how to derive the precise weightings from their judgements. Of course, the results of this approach may also be open to criticism of involving personal values, vested interests and bias.

Literature: as an alternative to relying on policy experts, the weighting values can be abstracted from the literature by reference to a respected study or studies. For instance, the weighting scheme used by the Development Report Card for the States (Corporation for Enterprise Development 1991) to combine different components into indices could be used as a basis for a single index of economic regeneration. However, it is unlikely that there will be a pre-existing study that covers exactly the same key issues as have been identified for another particular policy. Moreover, these weightings would still need to be expressed in a set of numerical values for each of the indicators generated.

Public opinion: a survey of the relative importance of the issues concerned may provide an objective measure of the public's overall views. For example, Rogerson *et al.* (1989) conducted a large-scale survey to gauge public opinion over the factors that make up the 'quality of life' in an area. It is, nevertheless, very unlikely that such weightings could be obtained off the shelf from an earlier study to match the requirements of a study undertaken for a different purpose. Due to the time and expense involved, conducting a new opinion survey may not be a practical option. The most ambitious attempt to use public opinion surveys to guide public policy was probably the Continuous National Survey in the USA during the early 1970s, which failed to gain the federal support needed to sustain the necessary constant updating and customisation (Rich 1981). The problem of unreliability of opinion polls also casts doubt on the adoption of this approach.

analysing the indicators in the Town and City Indicators Database study (Wong et al. 2004), it was clear that an area's performance on different urban visions may vary and move in different directions; the use of a composite index would simply conceal the inherent tensions and conflicts between different visions. Another issue of creating composite indices is the tendency of lending themselves to the development of league tables. This no doubt stimulates a lot of debate and media attention, but can also cause misrepresentation and be subject to distorted interpretation.

The ranking of a composite index may, nevertheless, be a very useful tool to start off an analytical process of studying the spatial distribution patterns of the phenomenon concerned. This aspect is illustrated in my work on analysing different pathways of local economic development (LED) trajectory (Wong 2002a). Based on a conceptual framework of eleven factors that are widely perceived to be the major determinants of LED, a set of twenty-nine indicators were identified to measure these factors. Principal component analysis was first used to examine the structure of relationships among the compiled LED indicators for local authority districts in England and to explore the spatial patterns that emerge from the analysis. The initial analysis started off with the ranking of each principal component to detect the spatial distribution patterns (for example, the first principal component is mapped in Figure 6.1). The findings of the analysis lend empirical support to the theoretical conceptualisation of different dimensions of LED. While acknowledging the useful function of rankings, my only reservation is that rankings on their own without further elaboration or explanation will not improve our knowledge of the issues concerned and at times may send out the very negative signal of being merely a numbering exercise. It is also important to recognise that rankings do serve a hard and fast policy function, for example, successive deprivation indices developed by the ODPM and its predecessors were used as the mechanism for regeneration funding allocation.

SIMPLIFICATION AND STRUCTURE OF ANALYSIS

The discussion above suggests that the default option of using composite indices is not necessarily a perfect solution of summarising information from indicator sets. There has been an increasing trend of seeking other methods to simplify the structure of indicators by using headline/flagship indicators, linking indicators in bundles, applying summary score systems and using multi-dimensional presentation methods.

The use of headline indicators offers a middle ground to balance the need of providing a manageable amount of information and the need of giving out useful details to inform policy action. For instance, there is a tiered structure of sustainability indicators in the UK (DEFRA website): with 147 sustainable development indicators and a subset of fifteen headline indicators (see Table 6.3). There is no definite approach regarding the selection of headline indicators. The selection can be based on subjective judgement of the importance of the issue concerned, or based on correlation analysis to choose an indicator that strongly correlates with the other indicators measuring the same aspect of performance.

HEADLINE INDICATORS OF SUSTAINABLE DEVELOPMENT

Another method is to use indicator bundles to link a small number of separate indicators into groupings to reflect different aspects of the social phenomenon concerned. Indicators within the bundle will be used in conjunction to explain a specific set of circumstances in relation to that particular aspect of the phenomenon. This method was used by Dunn et al. (1998) to develop indicators of rural disadvantage for the Rural Development Commission. For example, three indicator bundles – access to employment, quality of employment and the vulnerability of employment in the local economy – were used to measure different aspects of the local labour market. Each bundle consisted of a small number of indicators. The number of people affected by each indicator was identified for each ward and then summed up to produce a notional number of people affected within that particular indicator bundle. The total figures were expressed as a proportion and the indicator bundle value was then used to produce ranking as well as for further analysis.

An alternative method used to summarise indicator values is by reducing the value of indicators into a small range of scores (e.g. 1–4) and then adding up the scores across all indicators to produce a summary score for that particular aspect of the phenomenon concerned. This method was used by Ernst and Young (1998) to develop regional competitiveness rankings and to benchmark the performance of the Midlands against other European regions. The aggregate scores of skills, innovation, productivity and other performance were used to produce an overall 'regional competitiveness ranking'. Copus and Crabtree (1996) also used similar methods to examine socio-economic sustainability. They assigned a positive or negative sign to each indicator within the nine dimensions of socio-economic sustainability. A summary score of each dimension was then calculated. The advantage of this method is that it can handle both quantitative and qualitative indicators, and reduce a set of incompatible information into a simple set of scores. The assignment of scores (especially for soft indicators) to individual indicators and the production of summary scores will inevitably involve a certain amount of subjective judgement and weightings from the analyst. Hence, Copus and Crabtree (1996) commented that it would

Table 6.2 Statistical weighting methods

Z-scores: each variable is transformed into a standard form so that it has a mean value equal to zero and a standard deviation equal to one. The standardised score of each indicator for each area is then either added or subtracted, depending upon the interpretation of positive values. The biggest advantage of this form of composite score is its simplicity; it can be easily understood. It also allows policy targeting by ranking areas at a variety of spatial levels. This method, however, tends to oversimplify the data by ignoring the complex relationships between the issues which the indicators represent. It easily leads to the danger of 'double counting' when some indicators are highly correlated. Hence, this method is less appropriate for handling a large number of indicators.

Regression analysis: is a statistical model to provide a convenient summary of the importance of various indicators (independent variables) according to their strengths in explaining the variation of a single all-important measure (dependent variable). For example, Coombes and Raybould (1989) used VAT data to model factors affecting local enterprise activities. The advantage of regression analysis is that it could be used for description of the dataset analysed as well as predicting the outcome of the wider population or at a different spatial scale. The regression coefficient of each independent variable provides an automatic weighting on the dependent variable that they seek to explain. The biggest problem of this method is finding a single valid variable to represent the concept in a suitably rounded way. Ideally, the choice of the variables used in the model should be theory driven. In reality, in most cases, they are purely based on the past experience and knowledge of the analyst who performs the modelling task. There are also limitations of using regression models for prediction as it makes the assumption that the current model is still valid for the predicted observations.

Factor analysis: is used to identify a relatively small number of factors that can be used to represent relationships among sets of many variables. This is achieved by explaining as much variance among the variables as possible. Duguid and Grant (1983) used factor analysis to combine several indicators into a single deprivation score to prioritise areas of special need in Scotland. This approach was also used in the latest Index of Multiple Deprivation to combine indicators within the housing, health, education and accessibility domains (DETR 2000b). One of the strengths of this technique is that the obtained factor(s) help clarify the general concept on the basis of the empirical links within a set of indicators. It also provides an automatic statistical weighting of each variable on the factors. Hence, the factor score obtained for each area can be used for ranking. In the LED indicator study, Wong (2002a) used principal component analysis to examine the structure of relationships among the compiled LED indicators for local authority districts in England and to explore the spatial patterns that emerge from the analysis. Factor analysis, nonetheless, can be seen to have some disadvantages. The application of factor analysis involves critical decisions, such as which statistical options should be used in the statistical procedures, and which and how many factor(s) should be used for ranking. The process of assigning a label to each factor according to their attributes is also highly subjective.

Multi-criteria analysis: the results from a multiple factor analysis cannot yield a single ranking solution on their own. However, they can provide the basis for multi-criteria analysis. The factor scores of the chosen factors for each spatial unit can be assessed to see which exceed a threshold value on a set number of factors qualified. Massam (1993) described several versions of this method to illustrate the ways in which 'spatial coincidence' of several factors can contribute to policy-related analyses and decision support. The strength of this method is that it can be closely linked to policy concerns through, for instance, distinguishing which areas score highly on which particular factors. However, the operation of this method requires lengthy and complex explanation; no simple ranking can be calculated for the individual spatial units, although it is possible to rank an upper tier set of areas (e.g. local authority district) on the basis of the proportion of their population which live within those lower areas (e.g. Census wards) that fall into the target categories.

Cluster analysis: is a statistical technique that aims to classify areas into relatively homogeneous groups. This method has been widely applied in the private sector to create geodemographic classification schemes as a means of discriminating variations in consumer behaviour (see Batey and Brown 1995). The characteristics of each cluster can be identified from the descriptive statistics of each variable. This method can provide a very parsimonious solution by identifying the target areas in just a few clusters. It takes into account the different dimensions of the issues concerned within the classification process. Equally, there are notable disadvantages of cluster analysis as it requires detailed and debatable operational decisions throughout the whole statistical procedure. First, the measurement of some form of association or similarity between the areas is needed in order to show how many different groupings really exist in the study. It is then up to individual researchers to determine how many outcome groups they would like to obtain. The next step involves the profiling of the areas in order to determine their composition and to facilitate the labelling of each cluster. Also, the identification of which clusters should be considered to be the target' areas for a particular policy is based on the judgement of the policy-makers with respect to the characteristics exhibited by different clusters. No ranking of individual areas can be obtained as an area is either 'in' or 'out' of the chosen cluster(s).

not be sensible to compound such subjectivity by combining them into a single overall index.

Instead of reducing indicator values into a simple index or a summary score, one approach is to use multi-dimensional diagrams to present the indicator value. Herman *et al.* (1988) used 'snowflake' diagrams to illustrate the value of nine indicators they used to analyse the dynamic characteristics of US cities. The value of indicators, standardised against the overall average, was plotted along the respective axis assigned for each indicator in the polygon for each individual city. Similar graphic presentation was employed by the Asian Development Bank (Westfall and de Villa 2001) to develop city

Table 6.3 Department of the Environment, Food and Rural Affairs (DEFRA)

Objective 1: maintenance of high and stable levels of economic growth and employment	
Economic output	* total output of the economy (GDP)
Investment	* total and social investment as a percentage of GDP
Employment	* people of working age who are in work
Objective 2: social progress that recognises the needs of everyone	
Poverty and social exclusion	* indicators of success in tackling poverty and social exclusion
Education	* qualifications at age 19
Health	* expected years of healthy life
Housing	* housing condition
Crime	* level of crime
Objective 3: effective protection of the environment	
Climate change	* emissions of greenhouse gases
Air quality	* days when air pollution is moderate or high
Road traffic	* road traffic
River quality	* rivers of good or fair quality
Wildlife	* populations of wild birds
Land use	* new homes built on previously developed land
Objective 4: prudent use of natural resources	
Waste	* household waste, all arisings and management

Source: DEFRA (2004), sustainability indicators web page (www.sustainable development. gov. uk/indicators/index.htm).

holograms to allow comparison of a range of cities over a wide spectrum of indicators. These diagrams provide powerful graphical presentations of the growth and development of the cities and can also be used to compare the performance of different cities or to examine changes of each city over time. Figure 6.2 illustrates the use of such multi-dimensional diagrams to represent different indicator values of some large urban areas in England.

Besides the need to develop a parsimonious structure to handle the indicators, the approaches used to examine issues at different spatial scales and across different policy sectors also bear significant influence on the overall analytical framework. Since a large amount of data from a whole variety of sources is

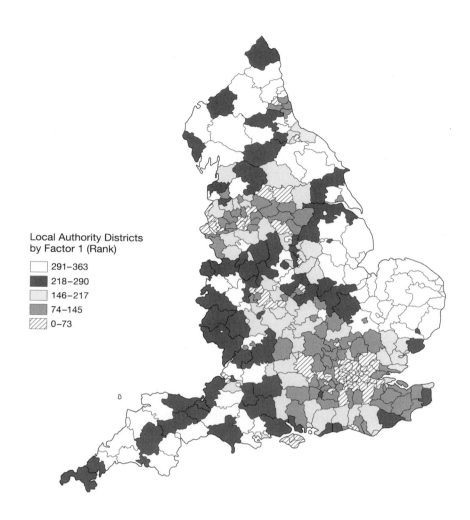

Figure 6.1 'Big city syndrome' (Principal Component 1)
Source: Wong, 2002a: 1850.

expected to be collected and assembled centrally to form a consolidated indicator database, the data has to be collected on a common spatial and temporal
basis, under a clearly identified set of definitions to allow meaningful analysis. This
is especially important if the analysis is an ongoing process of tracking patterns
of change of a social phenomenon. Due to the complexity of many social
phenomena, different issues are found more appropriately dealt with at different
spatial scales. Hence, a multi-spatial framework or spatial nesting is often used to

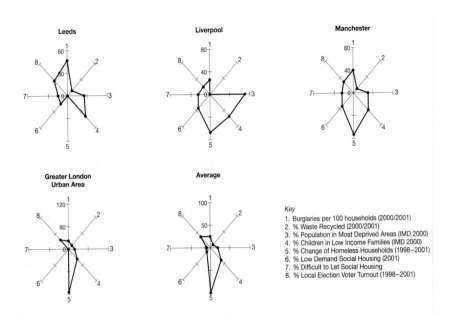

Figure 6.2 Multi-dimensional radar charts

provide a flexible analytical structure for the study of the social phenomenon concerned. Spatial nesting is closely related to the compatibility issue of data aggregation as well as the approaches used to develop and interpret indicators.

Robson *et al*.'s (1995) 1991 Index of Local Conditions serves as a good example to illustrate the use of a spatial hierarchical structure to develop indicator sets. The indicator value was built up from the smaller area (i.e. enumeration districts and wards) to the larger area (i.e. districts), with larger-area values being aggregates of smaller-area values. They also used a matrix approach to present the final index ranking for each district together with three other measures: the spatial *extent* of deprivation at ward and enumeration district levels, and the *intensity* of deprivation. The implementation of such a neat spatial structure is, however, heavily dependent on the availability of data at very fine spatial scales. More importantly, such a nesting structure of measurement may not be suitable for the exploration of certain social phenomena. Recently, there is a suggestion that indicators should be developed for different purposes at different spatial levels of concern. For instance, Kearns and Forrest (2000) provided a systematic account of the different constituent components of social cohesion and how they should be pursued by policy actions at different spatial scales (inter-city, city, city–region and neighbourhood levels). Table 6.4 shows their analysis on the interconnection between different social cohesion domains and spatial scales.

Sharing the same standpoint of Kearns and Forrest on the importance of addressing issues with different types of policy action, Innes and Booher (2000) proposed a three-tier indicator system to provide intelligence of city performance. These three types of indicators are system performance, policy and rapid feedback indicators:

> System performance indicators: a few key measures that reflect the central values of concern to those in the city and how the urban system is working.
>
> Policy and programme indicators: reflect the activities and outcomes of various elements of the system to provide feedbacks to policy-makers on how specific programmes and policies are working.
>
> Rapid feedback indicators: provide rapid feedback data to help individuals, agencies and businesses to make day-to-day decisions.

The European Commission (2000a) also proposed a similar reference framework for the monitoring of its Structural Funds. Indicators are collected to monitor three tiers of programme objectives: global, specific and operational objectives. It is interesting to note that both classifications rest upon a layered indicator structure of measurement by developing indicators from a general–strategic level to gauge the overall health of the urban system, through the measurement of policy outcomes, to the more imminent/intermediate measures of policy feedback. This approach of nesting indicator sets has the obvious advantage of serving different analytical and policy purposes, and to avoid information overload.

As explained earlier in Chapter 5, when developing the methodological framework for the Town and City Indicators Database (TCID) research, one of the concerns was how to strike a balance between avoiding information over-

Table 6.4 Addressing social cohesion at different spatial scales

Social cohesion dimension	National/ inter-urban	City/city-region	Neighbourhood
Common values/civic culture	**	**	*
Social order/social control		**	**
Social solidarity/wealth disparities	**	*	
Social networks/social capital		*	**
Place attachment/identity	*	**	*

Notes: ** domain in which urban governance attention and efforts are clearly evident. * domain in which there is a case for greater attention from urban governance.
Source: Kearns and Forrest 2000: 1003.

load and providing sufficient policy intelligence. In order to simplify the structure of the indicators used, a two-tier indicators system has been proposed. *Strategic* indicators are used to collect trend data on a small number of performance indicators that have been widely used by researchers to gauge urban change. Trend data provide a full picture of the condition of urban areas brought about by the process of socio-economic restructuring. These indicators measure both intensity and dynamics, and examples include population level and change, employment level and change, unemployment level and change and duration, and gross domestic product per head and change. The lower tier of the indicator system, the *vision* indicators, focuses on dealing with domain-based issues guided by the underlying conceptual framework of the five Urban White Paper visions, that is: 1) people shaping the future; 2) attractive towns and cities; 3) enhanced environmental sustainability; 4) creating and sharing prosperity; and 5) providing quality services.

In addition, a spatial hierarchical structure is proposed for the longer-term development of the TCID. There are three different potential scales of analysis: supra-urban areas, urban areas and sub-urban areas. Due to the constraint of data availability and the problem of developing analytical spatial units rather than using administrative units, these ideas are yet to be further developed. It is likely that local authority districts will continue to be used as the building block to provide a close match to the definition of urban areas. There is clearly a trade-off between the amount of data available and the use of more appropriately defined spatial units. However, as many researchers point out, the use of local authority district boundaries has distorted the analysis of competitiveness due to the issue of over- and underbounding. This means that an effort should be made to move away from arbitrary administrative boundaries in the long run. The unit of analysis involved at the supra-urban area level tends to be more appropriate for dealing with issues that are of strategic nature. For meaningful analysis of some economic and social indicators, the concept of city–region has a lot to offer. However, this may involve a major analytical exercise. Likewise, travel to work areas will provide useful functional areas to capture activity flows, though the commuting flow data is over-reliant on the decennial Population Census.

At the micro-level, the definition of local neighbourhoods is more problematic as the boundary of a neighbourhood is not easy to pin down (Gowman 2000). While there have been numerous attempts to define neighbourhoods, no definition is entirely satisfactory as there is a lack of clarity of what a neighbourhood exactly entails (Stewart 2000). Following the move towards community planning in Scotland, there has been serious debate towards deriving sensible neighbourhood units to serve a lowest spatial level of policy implementation, service review and public participation. It is interesting to find that the National

Health Service in Scotland (NHS Scotland 2001) recently went through a legal test of whether it secured adequate provision of pharmaceutical services in the neighbourhoods under the 1995 Regulations. While there is no definition of the term neighbourhood, this does not mean that there has been no judicial guidance. Lord Justice Banks provided some useful guidance to the issue far back in 1932:

> I pass now to consider what is indicated by the expression 'neighbourhood'. In this connection it is impossible to lay down any general rule. In country districts people are said to be neighbours, that is, to live in the same neighbourhood, who live many miles apart. The same cannot be said of dwellers in a town where a single street or a single square may constitute a neighbourhood.... Again, physical conditions may determine the boundary or boundaries of a neighbourhood, as, for instance, a range of hills, a river, a railway, or a line which separates a high class residential district from a district consisting only of artisans' or workmen's dwellings.
>
> (NHS Scotland 2001: para. 4.3)

The suggestion from the legal case is that neighbourhood must follow its ordinary meaning of what people may wish to consider where they live, think what is their neighbourhood, and what factors define that as being their neighbourhood. Unfortunately, this flexible proposition does not help in devising a universal neighbourhood framework for statistical analysis.

In spite of the difficulties, there is however a general consensus that neighbourhoods should not be entirely defined by existing administrative boundaries (Dunnell 2002). Having said that, in reality, the spatial definition of neighbourhood tends to link to some policy areas such as local strategic partnership areas and neighbourhood renewal areas. This causes the concern that, 'each policy, programme or initiative can define community or neighbourhood in a way that describes anything that the policy-makers want to promote at any particular time' (Stewart 2000: 3). Recent academic research has also been carried out to fine-tune the ecological approach of deriving neighbourhoods by employing a number of variables (e.g. density of social networks, institutional infrastructure, etc.) to account for the nature and context of social ties within the neighbourhood processes (Sampson 2004). This may offer some mileage for future development of the TCID.

In addition to the problem of defining neighbourhood, attention should also be paid to the transitional area between the commercial hub of the city and the suburban residential neighbourhood as such spatial areas are widely found to be derelict and dysfunctional. Again, there is no easy way to define such areas without making reference to local circumstances.

DATA ANALYSIS AND INTERPRETATION ISSUES

The production of an indicator database does not offer any knowledge or intelligence to inform policy action. Simple analysis such as ranking of indicator values can also be counter-productive as it is likely to be subject to all sorts of misinterpretation, which tends to be made out of context. Collecting data and constructing indicators are very cumbersome and laborious tasks; it is thus important to maximise the utility of such indicators to benefit public decision-making. To do so, value-added activities such as rigorous analysis and interpretation will be required. The discussion here focuses on the issues surrounding data analysis and interpretation.

CONNECTION BETWEEN THEORY AND MEASUREMENT

One of the obvious reasons of analysing indicators is to explore their patterns of spatial distribution and to investigate the degree to which there is measurable co-variation and interactive effect across different issues of the phenomenon studied. Such analysis tends to be guided by existing theories on the causal relationship between different factors or an *a priori* conceptual framework. There are, however, some genuine problems of connecting theory and measurement in practice.

First of all, the relationship between input factors and outcome phenomena is not easily identifiable as it could be non-linear and ambiguous. For example, although numerous academic studies suggest that the level of entrepreneurial activity is positively related to the quality of living of an area, the initial attractiveness of growing agglomeration economies, however, will soon turn out to suffer from the negative impacts of growth in terms of a deteriorating quality of life or a rising cost of living (Myers 1988). Castells and Hall (1994), for instance, documented the declining quality of living in Silicon Valley after its economic success – negative factors such as heavy traffic congestion, rising level of pollution from a so-called 'clean' industry and unaffordable housing prices have become the 'dark side of the chip'. This example serves to illustrate that there is not a linear, but probably an inverted U-shape, relationship between quality of life and economic performance. Even the concept of social inclusion, theoretically, can be a positive or a negative force depending on the scale and intensity of social and community ties (Boddy 2002). It is also widely recognised that rapid economic growth and unemployment can co-exist within the same urban space (Wong *et al.* 2000). The ambiguity over the causal direction of explanation makes the assessment of urban performance a difficult task.

Second, it is not always easy to interpret indicator values, because many factors are found to have ambiguous relationships with the social phenomenon being measured. Interpretation is thus very much dependent on which particular dimension of the phenomenon is being studied. For example, long-distance commuting can be good for the economic performance of a city as it draws upon a wider pool of talents and skills to compete in the global market place. However, it can be negative in environmental terms as commuting causes a large volume of travel flows and air pollution. Similarly, it is difficult to say whether a buoyant housing market with rising house prices is good or bad.

The third area of concern is the timeframe required to establish the relationships between different activities. For example, it is not easy to identify an appropriate timeframe to relate the causal mechanism between inputs and outcomes of urban change that involve a very lengthy restructuring process. After using a number of indicators to analyse different dimensions of competitiveness in Glasgow and Edinburgh, Bailey et al. (2002: 155) conclude that measures of Gross Domestic Product (GDP) per capita, unemployment and employment rates, or stocks of vacant or derelict land are all affected by long-term processes and they probably tell us little about current performance or future competitiveness.

The connection between theory and measurement remains to be a major challenge to social scientists, as socio-economic phenomena tend to tangle in a web of relationships that are not easy to be analysed in a systematic manner. The polyvalence nature of social phenomena also means that the outcome of research tends to be fuzzy images rather than clear-cut black and white answers.

ANALYSIS AND POLICY INTELLIGENCE

As discussed earlier, there are alternative ways of simplifying the structure of indicator sets including the use of composite indices. However, the approaches used to develop composite indices tend to be either arbitrary or too complicated to be transparent. If there is not a specific need to make use of the indicators for funding allocation, it is more important to tease out the key signals or messages that have emerged from the analysis of the indicator set and to disseminate the findings in a clear and direct manner to offer ideas and insights for policy-making. One approach to facilitate such analysis is by linking a small number of separate indicators into groupings to reflect different aspects of the phenomenon being studied. Indicators within the bundle will be used in conjunction to explain a specific set of circumstances in relation to that particular aspect of the concept. This approach shall be called an 'analytical indicator bundle' method. Commentaries on the spatial patterns emerging from the indicator values within the bundle will form a mini-profile or vignette of the concept being measured. The indicator bundle method was used by Dunn et al. (1998) to develop indicators of rural disadvantage. The purpose of their bundling approach was, however, to produce a summary value for the bundle by adding up the number of people affected by each indicator within the bundle for each area (see Hodge *et al.* 2000). The summary value then represents the number of people who are experiencing a particular aspect of rural disadvantage. While this offers one of the neat solutions of simplifying the number of indicators to produce a summary value, it also reduces the analytical potential of the indicators. Based on their ideas, I reinterpret the method to develop a more 'analytical' oriented approach so as to bring out the analysis from the indicator bundle. The essence is to emphasise the merit of analytical rather than technical synthesis of indicators. This idea was first proposed for the development of the Town and City Indicators Database (TCID) project (Wong 2002c; Wong *et al.* 2004) and was recently recommended as one of the methods to analyse indicators for the monitoring of the Local Development Framework (see ODPM 2005b).

In order to illustrate the potential of using the analytical indicator bundle approach to carry out analyses, a set of journey to work indicators was analysed for a number of case study areas (Wong 2003). These indicators aim to explore the commuting patterns by examining urban areas as a home (i.e. on a residence basis) and a work location (i.e. on a workplace basis) (see Table 6.5). By plotting the indicator values of distance to work by home and work locations in scatter plots (see Figures 6.3 and 6.4), the range of commuting experience of the case study urban areas is revealed. Urban areas that had high proportions of residents and workforce commuting short journeys (the top right-hand quadrant

Table 6.5 Journey to work indicators

Case study urban areas				
	a	b	c	d
Gr. London	44.5	30.7	38.2	40.1
New Addington	25.4	19.3	51.1	23.2
Prestbury/Macclesfield	60.0	28.2	67.0	24.1
Sheffield	52.3	14.7	49.3	20.0
Swindon	78.2	11.1	65.6	20.1
Sunderland	64.9	18.1	62.7	15.0
Washington	43.5	24.7	52.8	22.4
West Midlands	53.9	18.0	49.0	24.2
England	52.1	27.1	52.1	27.1

Notes:
a % with journey to work of under 5 km - residence based, 1991.
b % with journey to work of over 10 km - residence based, 1991.
c % with journey to work of under 5 km - workplace based, 1991.
d % with journey to work of over 10 km - workplace based, 1991.
Source: Wong 2003: 272.

of Figure 6.3) are relatively sustainable and self-contained. Examples include the Sunderland Urban Area, the Swindon Urban Area and the Prestbury/Macclesfield Urban Area. At the opposite extreme, urban areas with a very small proportion of residents and workers travelling a short distance to work (the bottom left-hand quadrant of Figure 6.3) are characterised by more diverse and complex commuting flows. The Greater London Urban Area serves as a good example.

Urban areas in the bottom right-hand quadrant of Figure 6.3 tend to have a smaller proportion of residents but a large proportion of workers commuting short distances. This is because such urban areas tend to be residential areas that serve as the hinterland for other job centres and they do not provide a lot of local employment opportunities. Of the case study areas, New Addington falls into this category of areas, with Washington showing some tendencies towards this type. The opposite situation is found in areas on the top left-hand quadrant of Figure 6.3 where there is a large proportion of residents but a small propor-tion of workers commuting short distances. These areas tend to be major cities and urban centres serving a wider catchment area, such as Sheffield and the West Midlands Urban Areas.

The distribution of areas in Figure 6.4 to a certain extent is a mirror image of the distribution in Figure 6.3. The top right-hand quadrant includes areas that are cross-commuting areas (i.e. have a relatively high number of in- and out-flows).

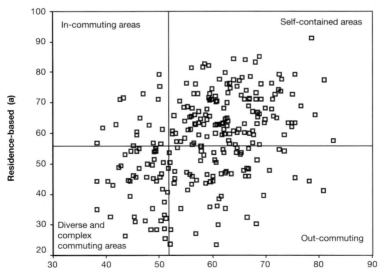

Figure 6.3 A scatter plot of short (under 5 km) distance journey to work–home vs workplace based values

Source: Wong, 2003: 273.

Examples include Greater London and the Prestbury/Macclesfield Urban Area. Areas in the bottom left-hand quadrant are more self-contained areas, such as Swindon and Sunderland, while the bottom right-hand quadrant includes urban areas that provide jobs for long-distance commuters as well as local residents, such as Sheffield and the West Midlands. Areas located in the top left-hand quadrant of Figure 6.4 are residential suburbs for outward commuters, such as New Addington and Washington.

The journey to work indicator bundle demonstrates that analysis can be enriched by putting a small number of interconnected indicators together to create a mini-profile of the issues concerned. If a composite index provides a technical/statistical synthesis of indicators, the bundle approach offers an analytical synthesis. The argument made here is that a more embedded analytical approach, supported by transparent methods and database, towards indicators development is a positive way forward to provide intelligence for policy monitoring and evaluation. Monitoring for quite some time was seen as negative feedback, a control exercise and an error-correcting mechanism to bring the plan back on track. This view should be encouraged to give way to a more positive and constructive view of monitoring – that is, a more forward-looking outcome rather than output-oriented approach with a broader focus given to the analysis.

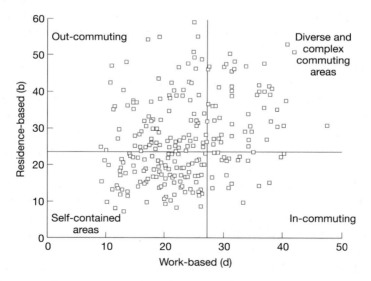

Figure 6.4 A scatter plot of long (over 10 km) distance journey to work–home vs workplace based values

Source: Wong 2003: 274.

TRACKING CHANGES: BENCHMARKING AND TREND ANALYSIS

In order to advocate the importance of eliciting intelligence from indicators, the analysis and interpretation process has to be an integral part of indicator development. Benchmarking with other comparator areas and longitudinal trend analysis are the two common approaches used to find out the nature and patterns of change of an area. Indicators have long been used to diagnose socio-economic change to inform and evaluate policy-making.

It is impossible to make a meaningful interpretation of an area's indicator values without making reference to other areas and to its wider spatial context. Interpreting policy implementation can be undertaken in absolute and relative terms. Whilst the analysis of indicator values and whether they are meeting set targets is a means of assessing implementation in absolute terms, it is useful to compare policy performance with the wider spatial context (e.g. the sub-region and the region) and other areas operating within a similar social, economic and environmental context. Such a benchmarking exercise helps the process of determining what is the very best policy and what standards should be set for policy targets. The concept of benchmarking was used in the management of industrial firms in the USA in the 1960s to look for best practices that led to superior performance (Camp 1989). The use of benchmarking was recently introduced into public administrations in the 1990s (Loomba and Johannessen 1997). In policy analysis term, benchmarking provides a yardstick to gauge the relative performance of an area by assessing its progress and achievement against other comparator areas or the wider regional and national level of change. This aims at identifying, learning and implementing the most effective practices to improve future performance. With the recognition that different areas perform under very diverse sets of socio-economic circumstances, there is a strong argument that the comparators should be selected on a like for like basis. Comparing areas with similar circumstances is especially important when operating under uncertain environments with a constant flux of unforeseen changes caused by external socio-economic forces. The use of benchmarking aims to assess relative performance, should it be economic growth under a buoyant economy or the resilience to economic decline during a period of recession, by taking into account their socio-economic circumstances as well as the external forces of change. This would help to reveal progress in real sense and to control the interruptive effects created by external events such as economic downturn. The awareness of the performance of others can help to explain how changes in these external conditions assist or hinder meeting policy targets.

It is, however, important to note that while benchmarking areas against each other and against future improvements is valid, this process can be exclusive and distort attention by focusing on particular negative aspects of urban problems and cause stereotyping (Taylor 2000). This suggests that when analysing indicators, one has to be sensitive to the presentation and dissemination of the findings. Area classification schemes created by the Office for National Statistics or other area typologies such as 'A Classification of Residential Neighbourhoods' can serve as useful tools for benchmarking areas on a like for like basis (Webber 1989). However, the typology used has to be appropriate to the issues being studied to provide a relevant context for interpretation. Robson *et al.* (2000: 13), for example, adopted the 1981 Office for Population Censuses and Surveys' (OPCS) area classification to explore urban change in the *State of the English Cities* report, as these area categories roughly illustrate the urban–rural continuum. In spite of the fact that the 1981 OPCS classification of local authorities was based on multivariate analysis of a range of 1981 Census variables, the value of the 1981 typology lies in its rough proximity to the urban hierarchy, which is useful to track counter-urbanisation of population and jobs (Breheny 1999: 225). Table 6.6 shows employment change between 1981 and 1996 by the 1981 OPCS area classification. Webber and Craig, who started the OPCS area classification, foresaw that 'there are advantages in sticking to one classification even if it is somewhat outdated ... because comparability is important and because, though out-of-date for some areas, the distinction still has some resemblance to current realities' (Webber and Craig 1976: 18).

Another important way of interpreting indicator values is exploring the trajectory, both direction and degree, of change over a period of time. Due to the descriptive nature of trend analysis, there has been a lack of academic interest

Table 6.6 1981–96 employment change by 1981 Office for Population Censuses and Surveys (OPCS) Areas

OPCS Area	1981	1996	Change	%
Inner London	2,023,741	1,915,496	-108,425	-5.3
Outer London	1,536,947	1,432,754	-104,193	-6.8
Principal cities	1,761,424	1,555,039	-183,272	-10.4
Other met. authorities	3,389,918	3,256,844	-28,769	-0.8
Non met. cities	2,758,206	2,821,178	62,972	2.3
Industrial areas	2,650,284	2,673,372	23,088	0.9
New towns	1,012,799	1,219,585	206,786	20.4
Resorts	1,050,562	1,126,562	76,000	7.2
Mixed urban–rural	3,216,695	3,849,955	633,260	19.7
Remote rural	1,893,335	2,178,486	285,151	15.1
Total	21,293,911	22,029,271	735,360	3.4

Source: Breheny, 1994: 204.

in exploring different aspects of urban change for a long time. In spite of the thorough review of the prospects of urban renaissance in the Urban Task Force Report (DETR 1999e), the spatial structure of employment distribution as a factor to development is neglected (Breheny 1999: 1). The awareness of the paucity of intelligence of these key trends of development has been heightened by a series of research projects commissioned by charitable organisations and learned societies. These include the two studies by the Town and Country Planning Association to examine the spatial distribution of population (Breheny and Hall 1996) and employment (Breheny 1999), a major Joseph Rowntree Foundation study of the job gaps in British cities (Turok and Edge 1999), and a study for the Council for the Protection of Rural England on the trends and reasons behind urban–rural migration over the last two decades (Champion et al. 1998). These studies provide very useful analyses of the geography of change and some of the key drivers behind the process; however, continuous research and trend analysis is needed to provide policy intelligence.

One issue concerning the approach used to conduct trend analysis is the timeframe of study. After reviewing a wide range of literature, the answer is that there is no explicit rationale of the length of time used to allow reasonable observation. Indeed, no research study attempts to provide a theoretical explanation of the choice of the analytical period. The timeframe is simply dictated by the availability of data series that offer a consistent dataset to allow reliable and valid comparison. However, most trend analysis tends to cover at least a 20 to 30-year time period (e.g. Cheshire et al. 1986; Giannias 1999; Turok and Edge 1999). When identifying indicators to measure city competitiveness, Bailey et al. (2002) argue that static indicators such as GDP per capita and employment rate are more of a reflection of historic outcome. They, therefore, suggest that change measures in GDP or uptake of vacant land would be more sensitive indicators to measure recent performance.

USE OF SOFT AND QUALITATIVE INFORMATION

Indicators tend to be found epistemologically associated with empiricism and positivism, and there is a natural assumption that indicators are quantitative and operational measures. In reality, many socio-economic issues are not susceptible to quantification and are inherently difficult to measure as they are either qualitative in nature or the assessment is a matter of opinion or subjective judgement. For instance, in searching for appropriate indicators to measure the urban visions for the TCID, it was found that not many meaningful measures could be used to measure community participation. This partly relates to the fact that there is not a consistent register of voluntary and community groups in a single database for easy access, but, more importantly, it is difficult to quantify the amount and nature of participation. Another set of difficult to measure issues were those related to aesthetic quality and attractiveness of towns and cities as they were subject to public opinions and could only resort to the use of survey data. There is a need to recognise that whilst the bulk of the indicators in a database will be based on hard and quantitative data, there will be other issues that are more appropriate to be understood through the use of soft indicators and qualitative information to enrich the interpretation.

To sum up the discussion here, it is clear that analysis and interpretation of indicators should adhere to a set of analytical principles to maximise the potential value of the indicators collected and to minimise the amount of bias and distortion involved as far as possible. Reality, however, often falls short of ideal because of data availability constraints and because the relationship between different issues is non-linear. The motto of indicator analysis should, therefore, emphasise the importance of knowing the methodological limitations and not over-exaggerating the findings. After all, indicators are not 'exact science'; they only indicate and provide a useful lens to diagnose and highlight interesting patterns of development that merit further analysis and exploration.

INFLUENCE THE INFLUENTIALS: ACCESSIBLE PRESENTATION

Most indicator research studies aim at influencing policy-making; it is then obvious that their findings have to be published and widely disseminated. The politician, Shirley Williams, offered some useful advice from an insider's perspective of how such presentation should be conducted:

> Many Cabinet Ministers work fourteen or fifteen hours a day, and there simply is not time to plough through research, or to read much outside one's departmental papers. Therefore the researcher depends upon an efficient system for abstracting, probably in not more than half a page, the broad outlines of the research and the conclusions reached from it, together with a clear indication of the source. Ministers will then start saying, 'find out more for me on this, tell me more about this', and, sometimes, 'I would like to read this, put it in the weekend box'. But without such abstracts large amounts of research simply slip below the horizon of the very policy-makers they are intended to reach.

She carried on to suggest that,

> purely from the point of view of somebody who has been at the receiving end of policy research studies. A lot of language is impenetrable and is becoming more so as people in the social science fields feel they must adopt a private professional

language similar to that in which many economists have wrapped themselves up for the last twenty years.... So language ought to be clear. The findings of the research, and recommendations arising from it should be clearly shown and separated from the text. There ought to be a summary of recommendations with reference back to the text in case people want to pursue the grounds upon which the recommendations were made.

Williams (1980: 5–6)

Complaints on the jargon used by academics in writing up their research findings were also reported in a recent study on the role of research in modernising local government (Sanderson et al. 2001). While such advice on the importance of writing research in a succinct and accessible manner is wise and sensible, those who frequently carry out policy research for the government will know that it is not an easy task to reduce everything into summary forms without misrepresenting or twisting the overall spirit of the research findings. Due to the multi-dimensional nature of indicator studies, the easy way out of reporting indicator values is going down the route of using composite indices. This, however, does not necessarily enlighten policy-making as complex issues are boiled down to a set of numbers or rankings. There are inherent tensions between the expectations of policy-makers and the delivery of academics: with the former expecting some straightforward black and white answers, and the latter offering a grey variety of findings with lots of caveats and caution notes. To be able to achieve what Shirley Williams suggested will require researchers to be brave and become more involved in policy issues and debates beyond the conducting of research. More importantly, personal judgement of the researcher in this give and take process becomes very critical.

Another observation is about the use of diagrams and maps to complement commentaries in reporting research findings. On the whole, the use of visual aides offers more accessible forms of presentation. However, it is interesting to note that some of the non-technical audience may be able to cope with two-dimensional diagrams very well, but would find any presentation that is multi-dimensional daunting. For instance, some indicators included in the TCID were presented in radar charts (similar to those presented in Figure 6.2) and the case study participants consistently found them difficult to understand. This is somewhat unfortunate as the real world does not function on a two-dimensional basis and a series of simple diagrams can only offer a glimpse of the picture and, in many ways, a disjointed one. Maps are interesting devices as they can elicit very extreme reactions. Through the case study of the TCID, it was found that some participants were very keen to see maps while others were totally disorientated. This shows that certain visual aides will require a particular aptitude to comprehend the messages and may not be that well suited for public dissemination and mass communication. Ironically, this means that jargon-free text remains as one of the most powerful tools in reporting findings. Diagrams, tables and maps should, however, be used sparingly in their simplest forms to complement the commentaries.

CHAPTER 7

METHODOLOGICAL PROCESS OF INDICATOR DEVELOPMENT: A REVISIT

PROCESS OF INDICATOR DEVELOPMENT

With the fetish of using evidence and statistical measures in the public policy arena, the green shoots of all sorts of indicator sets are sprouting. While there is a gust of enthusiasm towards this new indicators movement, this euphony can easily turn into a haphazard collection of statistics without any real sense of direction and purpose. The measurement of many of these abstract concepts is not underpinned by theoretically sound or policy-focused frameworks (Innes and Booher 2000; Sawicki 2002; Wong 2000). In order to avoid going down the fateful 'garbage in, garbage out' approach as witnessed in the past, there is a need to revisit some of the methodological issues involved in the process of indicator development. This chapter, therefore, aims to highlight the key components and provide an overview of the methodological process of indicator development.

Since the social indicators movement, different suggestions have been made over the process of developing indicators. The discussion here has largely drawn upon the four-step methodological framework (that is, conceptual consolidation, analytical structuring, identification of indicators and creation of an index) put forward by Coombes and Wong (1994). There are, of course, other ways to classify the process of indicator construction, for example, Zapf's (1981) six-step approach; nevertheless, the number of steps involved is often a variant of splitting or combining the four steps rather than offering any substantive new ideas. With the changing policy agenda and recent development in data infrastructure and methodological approaches, it is the right time to revisit the four methodological steps of indicator development. The arguments put forward here aim to fine-tune and adjust the original propositions to reflect on some new thoughts emerged from the cumulative experience of indicator research carried out by others and myself over the last decade.

The four-step approach is retained largely because it epitomises the beauty of simplicity to pinpoint the key milestones involved in developing indicators. This methodological approach has recently been employed by Hemphill *et al.* (2004a) to develop indicators for measuring sustainable urban regeneration. The four-step procedure (see Figure 7.1), working from general

Step 1: Conceptual consolidation
Clarifying the basic concept to be represented by the analysis
⇓
Step 2: Analytical structuring
Providing an analytical framework within which indicators will be collated and analysed
⇓
Step 3: Identification of indicators
Translation of key factors identified in Step 2 into specific measurable indicators
⇓
Step 4: Synthesis of indicator values
Synthesising the identified indicators into composite index/indices or into analytical summary

Figure 7.1 The four-step methodological framework of indicator development

to specific, is proposed as the basis for a consistent development process of indicators.

The final step is now renamed as 'Synthesis of indicator values' rather than 'Creation of an index' to provide a more accurate reflection of recent research ideas as discussed in Chapter 6. The pillars that form the backbone of this four-step framework are policy contexts, theoretical perspectives and methodological issues. The basic principles of each of these four steps are re-examined and discussed in turn in the following sections to provide answers to two key questions: What methodological steps should be involved in the indicator development process? And what detailed specifications should be made in each of these steps?

CONCEPTUAL CONSOLIDATION: THEORY AND POLICY

The first, and probably the most important, step to start off the process of indicator research is to clarify the basic concept that is to be represented by the analysis and to pinpoint the policy context and rationale against which the indicators will be used. Many of the key terms in policy discourses (such as 'deprivation', 'development potential' and 'quality of life') are subject to numerous interpretations; hence, it is essential to clarify the content of any such concept to facilitate subsequent analysis and to avoid any attempt to create a multivariate index by simply combining a haphazard choice of possibly related statistics without any theoretical basis. This is especially important if the eventual index is to be widely accepted as policy-relevant information (Ward 1980). The recognition of the basic conception is very important as it will lead to different indicator systems that represent different interests (Duvall and Shamir 1981). In a review of the best practice of area deprivation measures commissioned by the Joseph Rowntree Foundation, Lee *et al.* (1995) expressed concern over the confusion caused by the unclear definition and the indiscriminate use of different terminologies in studying deprivation. In the report, they commented that:

> The use of the same terms (social or spatial deprivation) gives the impression that different analyses are concerned with the same thing and that there is an agreed definition. However such definitions are often unclear. Moreover, where studies are variously concerned with deprivation, disadvantage, social exclusion or poverty they are attempting to measure, estimate and describe different phenomena. These labels are not always inter-changeable.
>
> (Lee *et al.* 1995: 13)

The Scottish Executive recently commissioned a project to develop a long-term strategy for measuring deprivation in Scotland. It is encouraging to see that the consultation document (Bailey *et al.* 2003) took conceptual clarification seriously and provided a detailed discussion over the distinction between 'deprivation', 'poverty', 'social exclusion', 'social cohesion' and 'social injustice'. Some commentators may view this kind of discussion too academic. This may appear to be so at first sight; nevertheless, clarity of ideas and thinking should be regarded as a prerequisite of any democratic dialogue and debate. Such an upfront attempt to deal with the definition of deprivation and to clarify the confusion of various terminologies helps to avoid any unnecessary misunderstanding in subsequent policy debates.

As discussed in Chapter 2, the normative nature of many indicator sets means that the architecture of the data has to be designed upon the needs of the policy client to serve a particular purpose. In order to have a good grasp of the policy context and rationale, interviews and discussion with policy-makers, stakeholders and end users will help to inform the focus of the study. It is, therefore, a fundamental task to address the basic questions of 'What is the purpose of the study?' 'What issues are linked to specific programme objectives (EC 2000a; ODPM 2002a)?' 'What policy instruments will be used?' 'And what is the appropriate spatial unit of analysis?' right from the very beginning of the study. These questions are vital in clarifying the issues that decision-makers consider to be most relevant, and in specifying the most appropriate spatial units for policy targeting that will then provide the underpinning of all statistical work in the later stages. An overall review of the best practice in related research in the policy arena should lead to a detailed discussion with the agencies involved. This process inevitably involves the value judgement of policy-makers who are the end users of the indicators. The experience (see Wong 1998a) gained

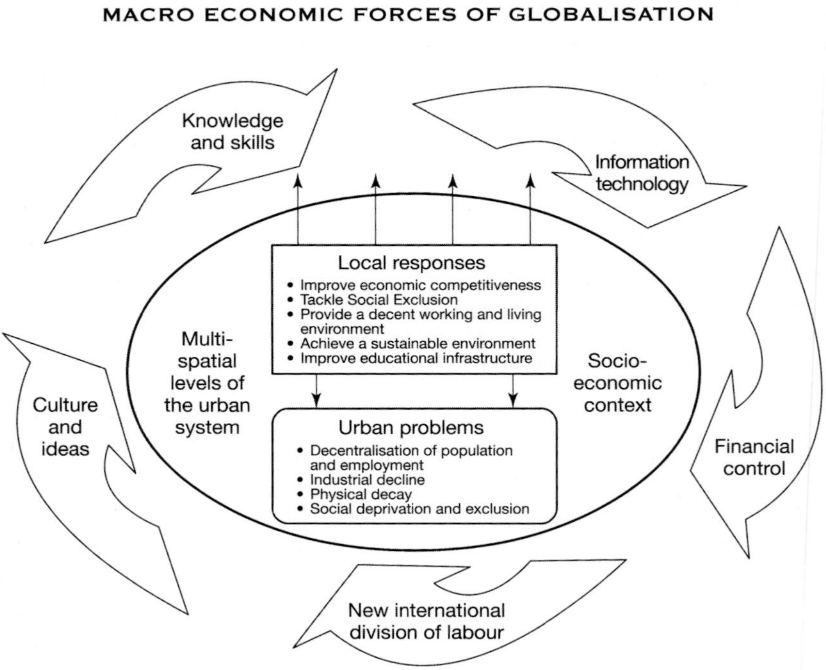

MACRO ECONOMIC FORCES OF GLOBALISATION

Figure 7.2 The dynamics of urban change
Source: ODPM 2004e: 3.

from the discussion with policy-makers over the application of the Local Economic Development (LED) indicators shows that the engagement of policy-makers in the process of indicators development can enhance the understanding of the policy operation environment and the subjective values and interests that policy-makers have over the research.

After the initial clarification of the basic concept and the policy context, further conceptual consolidation work has to be carried out to unfold the theoretical ideas that can be used to underpin the development of indicators. Since indicators tend to be surrogates of some abstract concepts, it is inevitable that they are associated with some kind of social and economic theories. Theoretical ideas can be elicited largely through a review of existing literature as well as from the views of experts in the field. In both the LED and the Town and City Indicators Database (TCID) studies (see ODPM 2002b; Wong 1998a, 2002c), similar issues arose when reviewing theories to develop an analytical framework. Both local economic development and urban change are a continuous process of spatial transformation. The driving force can be due to structural changes and historic inertia at the local level, as well as external factors from national and global forces, or indeed the interaction of internal and external factors. It is, however, impossible to separate the dynamic processes of change from the state of outcomes, as the outcomes themselves perpetuate further changes (see Figure 7.2). Concepts such as economic competitiveness, social exclusion and sustainable development all encapsulate the process of change as well as the state of performance. This makes the establishment of a causal input–outcome model of measurement difficult. It is also clear that current theoretical development cannot encompass the issues involved in its totality. Different academic studies tend to come up with somewhat different arguments as they all frame the research issues from a particular perspective and use different methods to ascertain the logistical relationship between input factors and outcome phenomena or between different phenomena.

The discussion here illustrates the problem as well as the importance of understanding the concept of measure. It is no easy task to find a consensual or a precise definition of many social concepts. While there is a sense of realism in acknowledging that the development of urban and regional theories will never be absolute and definitive, the choice and analysis of indicators should, as far as possible, be theoretically informed so as to avoid the development of a haphazard collection of statistics. More importantly, there is a fundamental need to clarify and delimit the meaning of the concept being measured to allow a common understanding of what is exactly the subject of measure.

ANALYTICAL STRUCTURING

After clarifying the key concept to be measured and the policy rationale against which the indicators are to be applied, the second step in the indicator development process is to develop an analytical framework. This framework aims to set out the structure and requirement upon which key elements/components of the indicators will later be developed and assessed. This analytical framework can be seen as the blueprint or the operational plan that provides a platform to underpin the collation (in Step 3) and analysis (in Step 4) of statistics.

There are two broad approaches to develop a framework of analysis. The 'bottom-up' approach mainly involves the listing of the key issues or factors that are considered to be important through, for example, a brainstorming session with experts or a literature search. This fast and pragmatic approach can easily fall into the trap of developing an incoherent set of indicators, as there is no objective criterion to cross-check the comprehensiveness of the list. In contrast, a 'top-down' approach starts from an *a priori* analysis of the concept concerned (i.e. Step 1), from which the causal relationship between different factors can then be derived to provide a framework of study (Coombes and Wong 1994). While recognising that a comprehensive account of the phenomenon may not necessarily be achievable in the later stage (for example due to the lack of appropriate data), the adoption of an analytical framework can ensure our knowledge of gaps and omissions.

The recent wave of the sustainability indicators movement has led to the development of a widely publicised conceptual framework of the *Pressure–State–Response* (PSR) model. This model classifies indicators according to their functions and roles in the decision-making process. Some sustainability indicators aim to provide a simple description of the current state of development (*state* indicators), while others are used to diagnose and gauge the process that will influence the state of progress towards sustainability (*pressure*, *process* or *control* indicators), or to assess the impact brought by policy changes (*target*, *response* or *performance* indicators). The sustainability indicator sets of the OECD and the United Nations are developed on the basis of a link model of 'pressure, state, response'. The PSR model provides a very neat and logical way of conceptualising the chain effect of human activities on the changing state of our environment and resources, and thus leading to the social and policy response to alleviate the pressure exerted on the environment. The operation of this model is, however, not that straightforward. When preparing for the *Indicators of Sustainable Development for the United Kingdom* report (DoE 1996), the Working Group abandoned the idea of adopting the model and opted for separating out indicators concerning the economy, the environment and the actors involved (Cannell *et al.* 1999). For others (e.g. Briggs *et al.* 1995; Dunn *et al.* 1998), the linear relationship captured in the PSR model is seen as oversimplifying the complexity of real life, and more complicated frameworks are thus proposed (e.g. Briggs *et al.* 1995; Post and Wieringa 1997). These extended models provide further sub-division of the process by chasing the driving forces outside the eco-system to identify different sources of pressure, and making a more fine-grained distinction between effects/impacts of change and the actions taken in response. However, the reality tends to be too complex to be captured by even these more elaborated models, as most effects also exert pressures on other variables (Dunn *et al.* 1998). This also echoes the tautological relationships found between different socio-economic variables over urban change in the Town and City Indicators Database (TCID) study.

Bearing in mind the earlier discussion that there are many untested assumptions of cause and effect in urban and regional theories which make it difficult to find a prudently proved causal framework to guide indicator analysis, the reliance on a top-down approach may not be plausible in some cases. This means that the choice of any input indicators could be arbitrary, and with limited or partial explanatory power. As far as current knowledge stands, it is fair to say that the development of environmental indicators is more advanced than that of socio-economic indicators in terms of moving towards a causal framework of analysis, though it is not without difficulties and reservations. Hence, there are obvious merits in combining both the top-down and bottom-up approaches to overcome the problem. The issues identified from bottom-up can be set against the top-down conceptual framework to test the comprehensiveness and validity of the issues included. The key components used to underpin the analytical framework can be derived from policy objectives, different dimensions of the concept involved (domain based), or different categories of issues that are relevant to the phenomenon studied. Although a comprehensive account of the phenomenon may not necessarily be achievable, the adoption of a clear conceptual framework can ensure our knowledge of errors and omissions, rather than perpetuate illusions of their absence (Duvall and Sharmir 1980; Gurr 1981).

When developing an analytical framework for the TCID study, a combined top-down and bottom-up framework was used. After an extensive review of literature, it became clear that the operation of different aspects of urban change may reinforce and enhance the restructuring process (such as quality of life and economic competitiveness), but their interaction can be contentious (the tension between achieving economic growth and a sustainable environment) or bear limited relationships to each other (such as the relationship between economic competitiveness and social exclusion). Due to the complexity and the intertwining of different socio-economic issues, it is impossible to untangle the web of input and outcome. In order to confront this dilemma, a structure–performance model (Carlisle 1972) was proposed to carry out analysis for the TCID study. This aims to highlight the differential socio-economic contexts against which the urban area operates and performs (Wong 2002c; Wong et al. 2004). Such an analytical framework does not aim to develop a causal model, but will provide a distinction between the more descriptive nature of the social conditions (*structure*) and the goal and outcome-oriented performance measures (*performance*).

Besides the overall structure of the analytical framework, there is a need to identify a set of analytical principles to guide the analysis and interpretation of the indicator set. For instance, six key principles were identified upfront to guide the development of the TCID project and to maximise the potential value of the indicators collected. These six principles also provide a framework for an overall evaluation of the key issues that emerged from the research study (Wong et al. 2004). These analytical principles are:

Tracking progress and change: the analysis of the indicators should provide clear narration of the nature and direction of change, and should not be subject to ambiguous interpretation.

Benchmarking and cross-comparison: it is sensible to interpret indicators by comparing areas with similar socio-economic conditions, in order to help to reveal progress.

Use of soft indicators and qualitative information: some of the issues involved may require the use of qualitative data and softer indicators to assess their progress or to enrich interpretation.

Exploration of co-variations and interactive effects: the indicator set should provide an opportunity in the longer term to investigate the degree to which there are measurable co-variations and interactive effects across different issues.

Consistency and comparability: a large amount of data from a variety of sources had to be collected and assembled centrally to form a consolidated database, so a common spatial and temporal basis is needed to allow meaningful analysis.

Multiple units of analysis: it is important to develop a multi-spatial framework to provide a flexible analytical structure for assessing the issues involved.

IDENTIFICATION OF INDICATORS

After establishing the conceptual and analytical groundings, the next step involves a laborious search for a wide range of possible indicators to measure the issues identified in the analytical framework. The drawing up of a 'wish list' of indicators is usually based on an extensive review of related policy practice and academic literature. In most cases, numerous potential indicators can be identified for each key issue. This is less true once data availability problems have been taken into account. This means a comprehensive search of statistical sources in the public domain, commercial databases and published directories in all relevant areas will be required. This task has become much easier over the last few years as many datasets can now be accessed online. This data-searching process will allow the assessment of information gaps in public statistical sources that affect the compilation of the dataset or a particular dimension of the dataset. Due to the polyvalence of many policy concepts, a single perfect indicator cannot usually be found to adequately represent each issue; the available data is more often in the form of proxy measures. This leads to a strategy of drawing upon a more broadly based set of measures in the analysis.

Since the measurement of indicators is very much a technical task, operational decisions have to be made in relation to the handling of statistics and other methodological issues. Knox (1978) identified various pitfalls in the construction of indicators, which include the difficulties encountered in the selection, availability and reliability of data, the problem of spatial aggregation of statistics, and problems of interpretation. Recognising the imperfection of the data means that the selection of indicators has to be rigorously assessed. Structured assessment of the value and practicalities of each potential indicator has to address five basic criteria: data availability, geographical specification (both coverage and spatial scales), time-series prospects, operation and implementation, and interpretation and relevance. Most of these issues have already been discussed in chapters 5 and 6. However, it is important to reiterate the fact that data availability, which frequently entwines with the problem of lacking fine-grained data, remains the fundamental millstone in restricting the eventual set of indicators developed.

Another major pitfall in indicator research is the lack of intellectual rigour in validating (see Carmines and Zeller 1979) and evaluating the measures. We have to question whether the indicators at hand are interpretable, relevant and adequately reflect the key issues of concern. For instance, the number of value-added tax (VAT) registrations is widely used to measure the enterprise culture of a locality. There are, however, different ways to express the value by dividing it either with the total number of economically active population or simply the total number of adult population. The problem of using the latter as the denominator is that it may give a slightly distorted picture of the level of business dynamics in areas where there is a higher concentration of pensioners, and will thus require caution in the interpretation of the indicator value. Hence, the recommendation is to use the total number of economically active population as a denominator to produce more sensible indicator values to compare the level of enterprise culture across different locations. Due to the difficulties in obtaining direct measures of certain factors, proxy measures are often used instead and more rigorous validity checks are therefore required. In other cases, the timeframe of data collection for the numerator and the denominator may vary and thus may affect the interpretation of indicator values. Using the VAT example again, the total number of new VAT registered businesses can be updated on an annual basis, but the total number of economically active population is based on estimates made at a particular point of time. It is then important to issue health warnings to guide the interpretation of such indicator values.

The implementation of the five suggested appraisal criteria should be carried out within a structured schema. In compiling the data for the TCID study, a total of 125 indicators for the five visions set out in the *Urban White Paper* (DETR 2000d) were identified and assessed by the research team. In order to ensure the quality of the TCID, potential indicators were assessed systematically by a set of appraisal criteria and the assessment was recorded in the baseline data information sheets, and the full set of information sheets was compiled and stored in a 'meta database'. An example of the assessment sheet for the indicator 'new homes built on previously developed land' is shown in Table 7.1. This example illustrates how this indicator proposed under Vision Two of the White Paper, 'attractive towns and cities that use space well', was evaluated. From the

Table 7.1 Data assessment sheet used in the Town and City Indicators Database Project

Name	New Homes Built on Previously Developed Land
Brief Description	The percentage of new dwellings built on previously developed land (i.e. housing recycle rate) in 1997–2000.
Indicator reference number	V 2.15
Type of data (e.g. Census, survey or administrative)	Administrative
Data collection source	ODPM Land Use Change Statistics, contact person: David Kelly Email: David.Kelly@odpm.gsi.gov.uk
Accessibility: permission, charges etc.	ODPM Land Use Change Statistics: free of charge on request
Input spatial unit: i.e. maximum spatial disaggregation	Unitary Authorities/Local Authority Districts
Output spatial units	Unitary Authorities/Local Authority Districts best fit to a sub-set of 78 Primary Urban Areas
Most up-to-date information	Land Use Change Statistics: 2001
Time-series	Land Use Change Statistics: 1985
Updatability	Annually
Format: Electronic/paper	Electronic – MS Excel
Implementation/Specification	Percentage of new housing completion on previously developed land, 1997–2000 average. When aggregating the percentage data from UA/LAD to urban areas, need to apply household weightings.
Interpretation	Percentage of new home built on previously developed land could be an indication of the attraction of urban areas. This indicator is also relevant to Vision 3 as the government's sustainable house planning policy sets out that at least 60% of all new housing should be built on previously developed land.
Health warnings	Land Use Change Data: due to the quality of some data, it is more reliable to take the average of several years' data. However, the 1997–2000 data are subject to a large margin of error, especially in the more rural authorities (changes in rural areas take up to five years to be detected by Ordnance Survey).
Potential solution(s)	It is more reliable to use the average value of the Land Use Change Data for trend analysis. There is a

	need to re-examine these data when further informa-tion is available from Ordnance Survey in the future.				
Key/further developments					
Urban Visions	1	2	3	4	5
Relevance score to Vision objectives	0	3	2	0	1

Note: 0 = not relevant, 1= minor relevance, 2 = partial relevance, 3= highly relevant.
Source: Wong et al. (2004): 106–7.

assessment, it is clear that ODPM's land use statistics offer a very relevant measure for monitoring Vision Two as well as for Vision Three of 'enhanced environmental sustainability'. The data is updated frequently and has been collected since 1985 to allow longer-term trend analysis, though there are health warnings attached to the reliability of some data. One of the major problems of using the housing data is its spatial scale of collection at the local authority district level, which is too crude to fit the definition of some urban areas. Similar assessment was carried out for all 125 prospective indicators, but only fifty-five were analysed (see Wong et al. 2004). The proposal to drop the other indicators was attributable to a number of reasons, including the lack of appropriate data sources, the indicator was too ambiguous for interpretation and/or the indicator was of minor or no relevance to the urban visions.

SYNTHESIS OF INDICATOR VALUES

The final step in the development of indicators involves the synthesis of the indicator values. One common practice is to develop a composite index by synthesising the proposed indicators, according to their relative importance, into a single measure that will be used for policy targeting. If a single most representative indicator can be identified for each key factor in the analytical framework developed in Step 2, the issue of weighting can simply concentrate on the relative importance of each factor without the need to consider indi-vidual indicators. However, practical problems such as data availability as discussed above usually impose constraints on the selection of indicators and their quality. It is, therefore, rarely possible to find indicators that can perfectly represent the key factors of the analytical framework. Because of this limita-tion, it is important to examine the properties and the reliability of individual indicators in the process of creating a combined index. For instance, the very similar statistical patterns exhibited by two indicators (which is expected to represent two different issues) may possibly imply one or both of these indica-tors are poor measures of the key factors concerned. Data validation is thus

Table 7.2 Types of statistical standardisation techniques

- Z-score: this method expresses the value of an indicator for a spatial unit as the number of standard deviations from the average of all spatial units concerned. This technique focuses on the proportional and relative size of the observed value in comparison with the overall average. This method was employed in the calculation of the 1981 Deprivation Index (DoE, 1983).

$$Za = (Xa - Ma) / SD$$

Where Z is the z-score for indicator 'a' in a given spatial unit, Xa is the percentage value of indicator 'a' for the spatial unit, Ma is the mean percentage value of indicator 'a' for all spatial units, and SD is the standard deviation of the indicator for all spatial units in the study.

- Chi-square: the calculation of the chi-square is based on absolute rather than relative values. This method compares the observed and the expected size of an indicator value against the overall average. The larger the chi-square value, the more statistically meaningful is the divergence between the observed and expected values. Signed chi-square method was employed by Robson *et al.* (1995) to develop the Index of Local Conditions for the DoE.

$$Ca = (Oa - Ea) (Oa - Ea) / Ea$$

Where Ca is the chi-square value for indicator 'a', Oa is the observed value for a and Ea is the expected value for a.

- Range: this method aims to devise a scoring scheme to ensure that each criterion contributed equally to the total score and can be expressed in a range of 0 to 100 (Little and Mabey, 1972). This technique, however, reflects neither the relative nor absolute size of the indicator values.

$$Ra = (Oa - La) / (Ha - La) * 100$$

Where R is the range score for indicator 'a', Oa is the observed value for a, La is the lowest observed value and Ha is the highest observed value.

Source: Lee *et al.*, 1995: 19.

considered to be a prerequisite before seeking a weighting method to create a composite index.

Before putting the index together, it is important to undertake a preliminary inspection of the statistical properties of indicators, such as their frequency distributions and the correlation coefficients between different indicators in the compiled database. In order to allow meaningful comparison of indicators, *standardisation* is commonly used to provide a consistent scale of measure to avoid the exaggerated influence of certain indicators. Unintended weighting effect of an indicator on a combined index can occur under two different situations: 1) when indicators are expressed in absolute numbers, adding up raw values together will put more emphasis on those indicators with larger absolute values; or 2) when indicators are expressed in percentage term, an indicator with a larger range of value will have a disproportionate effect upon the combined index than an indicator with a smaller range of value. The technique of standardisation is thus used to overcome this problem (see Table 7.2).

The initial data processing in some cases will also involve the application of *transformation* procedures such as log and square root transformations to normalise the indicator value. Transformation technique is applied to reduce the skewness of data values and to achieve a symmetrical bell-shape distribution of data. This will then allow the application of statistical techniques that are dependent on the assumption of a 'normal' distribution of data. In producing the 2000 Index of Multiple Deprivation (Noble *et al.* 2000b), transformation procedures were applied to data showing a strong degree of skewness. The downside of using complex transformation techniques is that they can dampen the degree of variations in data distribution when certain social phenomena are genuinely spatially discrete (Sammons *et al.* 1994). The decision of transforming indicator values can alter significantly the distribution of the more discriminating indicators and those that are more spatially polarised, for example the pattern of social deprivation observed by Lee *et al.* (1995). My personal view on transformation is that it should only be performed if certain statistical techniques based on the assumption of normal data distribution are required; otherwise, the indicator value should be presented as it is to allow the detection of the genuine spatial distribution patterns. It is also important to note that transformation tends to lose information that may be of importance to policy and add an extra layer of technical barrier to potential users' understanding of the index (Martin *et al.* 1994).

After the initial data processing of the indicator value, indicators can be combined together to create a composite index by applying a weighting scheme to individual indicators. As explained in Chapter 6, any recommendations on the method used to produce a combined index have to reflect the balance between simplicity, statistical robustness and flexibility. The honest answer is that there is rarely a simple 'right' or 'wrong' approach, but there are more appropriate solutions to handle a particular set of indicators to serve a particular analytical need. Another issue in relation to the development of composite indices is the problem of data reduction. Composite indices tend to lend themselves to the production of rankings and league tables. Although composite indices have the advantage of providing a hard and fast technical synthesis, they conceal detailed information on different aspects of the phenomenon studied. This is especially problematic when the relationship between the indicator and the phenomenon concerned is ambiguous. The reduction of an entire database into a single value will inevitably mask the tremendous analytical potential of the data. This leads to the suggestion of adopting a more analytical approach to present indicator values (Wong 2003) as explained in Chapter 6.

REMARKS

The discussion above aims to encapsulate the process involved in the creation of quantitative indicator sets. The proposed methodological steps serve as a norm of what indicator research should be achieved in methodological terms. However, the arguments put forward are also grounded on real-life policy research experience. This aims to inject a sense of realism to make sure that the proposed approach is methodologically sound as well as practically plausible to meet with policy needs.

PART III

CASE STUDIES

CHAPTER 8

DEPRIVATION INDICATORS

INTRODUCTION

Of the different types of policy indicators discussed in Part III of this book, the development of deprivation indicators can easily be rated as the most mature as well as the most politicised. This is partly related to their long history of development and partly related to their pragmatic value in informing resource allocation. As observed by Gordon, 'the construction of census based deprivation indices is one of the most economically important uses of social statistics since they form a key element in the allocation of both local government and health resources' (1995: S39). It is this political significance that has fostered a substantive volume of literature on the methodological approach used to develop deprivation indicators. Research has also been conducted to provide overviews of their development throughout the last two decades (e.g. Hayes 1986; Hirschfield 1989; Lee *et al.* 1995; Morris and Carstairs 1991; SEU 2000). It is, therefore, not possible to rehearse all these arguments at length here. The focus of this chapter is thus to identify the key trends of development in deprivation indicators and to highlight some of the key issues that intersect with their conceptualisation, methodology and policy use. The discussion of key issues will largely draw upon the range of deprivation indices developed by central government and others since the Department of the Environment's 1981 Deprivation Index (DoE 1983). Table 8.1 provides a list of key deprivation indices.

UNDERSTANDING DEPRIVATION

Since the publication of the Index of Deprivation by the DoE (1983), there have been numerous attempts to seek clarification of the concept of deprivation. Deprivation is not an easy-to-grasp concept, as its exact meaning is somewhat vague and ambiguous. This makes the development of a deprivation index a continuous challenge. As Greer (1969) pointed out, many fashionable policy concepts were embraced by policy-makers because of their vagueness. These concepts are then subject to a dynamic process of definition and redefinition, and become scientific problems for systematic

Table 8.1 Key deprivation indices since the 1980s

Author	Description of index, data source and standardisation method
DoE (1983)	The 1981 Index of Urban Deprivation for guiding expenditure under the Urban Programme, census data, Z-score.
Jarman (1983)	Underprivileged area score study for targeting primary health care resources, census data, Z-score.
Townsend (1987)	Material deprivation score (widely applied in poverty study and health inequality analysis), census data, Z-score.
Carstairs and Morris (1989)	The Scottish deprivation score for analysis of Scottish health data, census data, Z-score.
Forrest and Gordon (1993)	The construction of social and material deprivation indices following the release of 1991 Census, census data, indicator % is divided by the range.
Robson et al. (1995)	The 1991 Index of Local Conditions for the DoE, a mix of census and non-census data, Chi-square.
Gordon and Pantazis (1995)	The Breadline Britain score - following the method of Gordon and Forest (1993), weightings were derived from the Breadline Britain in the 1990s survey to produce a census deprivation index, census data, estimated % of poor.
DETR (1998b)	The 1998 Index of Local Deprivation was an update of the 1991 Index of Local Conditions, mainly non-census data with 3 census indicators, signed Chi-square.
Dunn et al. (1998)	Indicators of Rural Disadvantage were developed for the Rural Development Commission for developing and delivering policies and programmes that impact on rural areas, a mix of census and survey data, in % and ranking.
Noble et al. (2000b)	The Index of Multiple Deprivation 2000, make use of benefits, other administrative and census data, ranking of values.
Noble et al. (2004)	The Index of Multiple Deprivation 2004, make use of benefits, other administrative and census data, ranking of values.

Source: adapted and extended from Lee et al. (1995) and SEU (2000).

investigation, and the findings then feed through to the policy circle. Deprivation is surely one of these intriguing concepts that has been examined and re-examined at different times when new terminologies and jargons enter policy discourses. Deprivation itself was a new jargon when introduced to the policy circle in the 1980s. There has thus been continuous effort to distinguish deprivation from other similar concepts such as *poverty* and *disadvantage*. However, with the introduction and popularisation of new policy concepts like *social exclusion* by the European Commission and the government in the 1990s, attention has shifted to the distinction between deprivation and these related concepts (e.g. Bailey *et al.* 2003). The discussion here aims to provide a snapshot of our understanding of the concept of deprivation and how it is related to other similar policy discourses.

While some researchers such as Ferge (1987) held the view that the change of terminology between poverty and deprivation is a matter of convenience, others have tried to inject some clarity to the debate. Townsend (1987) provided one of the most authoritative definitions of deprivation. Based on his previous renowned work on poverty, he defined deprivation 'as a state of observable and demonstrable disadvantage relative to the local community or the wider society or nation to which an individual, family or group belongs' (Townsend 1987: 125). This definition focuses on conditions, should they be the physical, environmental and social states or circumstances, rather than purely resource based, as poverty is defined. Deprivation also refers to specific and not only general circumstances, and therefore, according to Townsend, can be distinguished from the concept of poverty. He further suggested that deprivation takes many different forms and a distinction can be drawn between 'material' and 'social' deprivation.

Other commentators such as Hirschfield (1989) and Saunders (1998) concurred that poverty is related to a lack of material resources, whereas deprivation relates to a broader denial of opportunities and an inability to participate fully in society. Hirschfield regarded poverty as an absolute concept, and deprivation as a relative one. Dunn *et al.* (1998), however, commented that, when measuring poverty by comparing income against a minimum standard of living or threshold, it became a relative concept too. While it is possible to semantically distinguish poverty from deprivation, the two concepts are tightly linked. The concept of poverty refers to a lack of income and other measures; it is nevertheless this lack of resource that makes the poor excluded from ordinary living patterns and it is highly unlikely that they can escape from these conditions (Gordon 1995: S40). With regard to the differentiation between deprivation and disadvantage, Brown (1983) regarded the latter as a more severe condition with damaging consequences extending beyond any immediate depriving effect.

Another major area of debate tends to surround the concept of multiple deprivation. The terms deprivation and multiple deprivation are used loosely, and often with little reference to their specific meanings (Gordon 1995). Multiple deprivation, at one level, can simply relate to the polyvalence nature and multiple dimensions of deprivation. However, it does contain a deeper meaning. In order to remove the conditions of deprivation, public resources should be distributed to those most in need – that is, those suffering from different forms of deprivation. As Holtermann (1975: 40) suggested, the ideal circumstances would be to find out whether there were substantial numbers of people suffering simultaneously from different kinds of deprivation. Unfortunately, due to the lack of personal and household level data to perform such a counting task, successive deprivation indices can only develop area-based measures to find out the spatial coincidence of social, economic and environmental disadvantages (Pacione 1995). All we can identify is the area where there is simultaneously a high proportion of households or persons suffering one kind of deprivation and a high proportion suffering another kind. The problem is that we never know the actual level of multiple deprivation because area measures cannot tell us whether it is the same individuals who have both kinds of deprivation. Due to the difficulty of finding suitable technical methods to measure the concept of multiple deprivation satisfactorily, there has been serious academic and policy debate over the value of using area-based as against to people-based regeneration policy to reach those who are most in need. This issue will be further explored in the next section.

In the 1990s the term 'social exclusion' entered the discourse through the European Commission, as it is more palatable than 'poverty' (Room 1995). This jargon was subsequently imported to the British policy circle as the Social Exclusion Unit was set up after the 1997 General Election within the Cabinet Office to tackle deprivation. The notion of social exclusion focuses primarily on relational issues and the denial of access to the major societal system such as the lack of social integration, inadequate social participation and lack of power (Room 1995: 105). In this sense, it is closer to the concept of disadvantage. According to Bailey et al. (2003: 10), both deprivation and social exclusion include material and social/relational dimensions; exclusion studies however tend to give greater weight to the latter and they tend to focus on the processes which lead to exclusion rather than the outcomes of these processes. It is, nevertheless, not an easy task to separate the outcomes of deprivation from the risks or conditions that may lead to deprivation (Coombes et al. 1995). The concept of deprivation may appear to emphasis outcomes rather than the process; in practice, this distinction does not hold. Each outcome of deprivation will in turn become a condition that traps the vulnerable into a vicious spiral of problems.

The discussion here simply shows that there is no consensus over the definition of these terminologies and one should not overstate their differences as there is a certain degree of overlap among them. However, the way deprivation is conceptualised bears important technical and political implications for the process of developing indicators and composite indices. This will be discussed in the next section.

CONCEPTUAL AND MEASUREMENT ISSUES

There has been longstanding debate over a number of issues, both conceptual and technical, which impinge on the development of deprivation indicators. Five key issues are discussed here as they illustrate the political sensitivity of the technical approaches used in developing deprivation indices. These issues include the relevance of indicators to urban and rural circumstances; the extent to which they are measuring deprived people or deprived places; whether vulnerable groups should be used as proxy measures; the dilemma of choosing appropriate spatial units of analysis; and the implication of using different weightings, standardisation and transformation procedures.

THE URBAN–RURAL DIMENSION

The widespread public debate on indicator measurement and usage was sparked by the publication of the DoE's 1981 Deprivation Index. The Index was originally aimed at measuring urban deprivation (DoE 1983), but was subsequently used to assess the relative levels of deprivation across all English local authorities, including both urban and rural authorities. The application of this index and its successors in all sorts of policy contexts had, in turn, discriminated against rural areas that had less chance of receiving government funding. This led to severe criticism from the rural community, as neither the indicators nor the methodology used were sensitive in identifying rural deprivation problems. The situation was exacerbated after the DoE published its 1991 Index of Local Conditions (Robson *et al.* 1995). The use of signed chi-square methodology to express the indicator value meant that it effectively reduced the weighting given to areas with small population size, hence the rural areas again lost out.

Based on the special tabulation of the 1981 Scottish Census data for enumeration districts, Knox (1985) found that there was a rural–urban dimension in deprivation terms. In urban areas, there were variations between local housing markets, with multiply disadvantaged households being more prevalent in public sector housing. Rural areas, on the other hand, were characterised by a significant element of the 'property-owning poor'. The nature of the problem in rural areas tended to be dominated by the combination of pensioner households of low socio-economic status, whereas those in urban areas tended to have low socio-economic status, unemployed, in poor health, with disability, living in over-crowding conditions and single-parent families.

There has been strong lobbying of government from local authorities in rural areas to revise the methodology used in the so-called official deprivation index. The then Rural Development Commission went ahead to commission

Cambridge University to develop a new approach to measure rural disadvantage (Dunn et al. 1998). The research team pointed out that there were two main areas of concern with these urban deprivation indices (Hodge et al. 2000: 1,871):

the variables selected have less relevance to the characteristics of rural disadvantage, and
the incidence of disadvantage is more diverse and dispersed in rural areas.

The crux of rural deprivation is very much related to the issue of accessibility (Coombes et al. 1993a). The importance of accessibility to public services and facilities in rural localities was explored in recent research (Dunn et al. 1998; Noble et al. 1999) and incorporated as one of the domains in the 2000 and 2004 English Indices of Multiple Deprivation (Noble et al. 2000b; 2004).

When the 2000 Index of Multiple Deprivation was released, there was a sense of relief from the Countryside Agency (the successor of the Rural Development Commission). This was explicit in its press release:

> The fact that rural deprivation exists and can be as bad as in some of the worst inner city areas, is recognised in a new government index of disadvantaged areas in England. More rural areas will now be able to benefit from area-targeted initiatives such as New Deal for Communities, Single Regeneration Budget and the Neighbourhood Renewal Fund.
>
> (Countryside Agency 2000)

The Countryside Agency subsequently commissioned the same research team at Oxford University to provide more detailed comparative analysis on the urban–rural dimension of the index, and was content with the approach and used the index for its policy targeting. However, academics such as Kearns et al. (2000: 1557) argue that a separate exercise that seeks to measure deprivation in rural areas, taking into account the effects of remoteness, transport costs, accessibility and poverty among those people in low-paid employment, should be undertaken.

AREA-BASED MEASURES AND AREA-BASED POLICIES

Due to the lack of data for individuals and households, deprivation indicators tend to be derived from Census small-area statistics, which are area-based information (Bailey et al. 2003). This raises a serious question over the extent to which the pattern of deprived and disadvantaged areas reflects that of disadvantaged households and individuals (Knox 1985: 414). This type of area analysis may be subject to the problem of ecological fallacy, that is, spurious spatial auto-correlation. That means areas of high levels of deprivation may have a high proportion of particular social groups, but it cannot be assumed that these groups are themselves deprived (Fieldhouse and Tye 1996). The use of a series of area-based indicators to gauge multiple deprivation is particularly problematic. This is because the spatial concentration of individual aspects of deprivation was found to be quite low (Holtermann 1975). In her seminal study, Holtermann found that even with severe overcrowding, an indicator that showed the highest levels of spatial concentration, the target of the worst 15 per cent of the enumerate districts could capture less than two-thirds of all households with this type of deprivation. Her work posed the challenging question of whether different deprivations generally occur within the same households in an area that is deemed as multiply deprived. The question asked by Holtermann nearly three decades ago remains unanswered. In the discussion of the latest 2004 Index of Multiple Deprivation, the research team dealt with this problem upfront:

> It is possible to look at single forms of deprivation at an area level and state that a certain proportion of the population experiences that deprivation or a proportion experiences some other forms of deprivation etc. and describe at an area level the combination of single deprivations as area level multiple deprivation. The approach used here conceptualises multiple deprivation as a composite of different dimensions or domains of deprivation. It, however, says little about the *individual* experience of *multiple* deprivation.
>
> (Noble et al. 2004: 10)

The crux of this debate is closely related to the political argument of whether the use of an area-based index for area-based policy (e.g. Pattie 2001; Joshi 2001) is effective in tackling deprivation issues. Area-based policies have been criticised as ineffective because the majority of deprived people do not live in deprived areas (Kleinman 1999). Some commentators thus argue that it would be more effective to have policies that target deprived individuals wherever they live (McCulloch 2001; PIU 2000b). Knox (1985) found that disadvantaged households were disproportionately concentrated in the worst-off areas and, in particular, multiply disadvantaged households. He found that 51 per cent of all households in the worst 1 per cent of enumeration districts in Scotland were multiple-disadvantaged, and thus supported the efficiency of area-based policies. He, however, warned that there was a considerable number of disadvantaged households living outside deprived areas as the worst 5 per cent of enumeration districts only accounted for 10.7 per cent of all multiply disadvantaged households in Scotland. More recently, the relationship between spatial scale and concentration of unemployment was also examined by Johnston *et al.* (2003), using a threshold profiling approach. They found that less than 10 per cent of all unemployed lived in areas with twice the national unemployment level when examined at district level; however, it moved to over 30 per cent when shifting the scale down to enumeration district level. This means that the focus of policy targeting improves when the spatial scale is smaller. Nonetheless, the researchers concluded that different aspects of disadvantage might require a different spatial scale of analysis.

By measuring deprivation at the level of the individual rather than that of the area, it will be possible to overcome the problems of ecological fallacy. Fieldhouse and Tye (1996) used the 2 per cent individual Sample of Anonymised Records from the 1991 Census data to ascertain the micro-level of deprivation. Their findings from using individual-level data showed, as conventional area-level analyses suggested, that multiple deprivation was heavily concentrated in particular social groups and in particular geographical areas though the strength of the relationships at the individual level was not as powerful as the aggregate area data might suggest.

The research findings basically point to the conclusion that using finely spatial scales in area-based indicators can improve the effectiveness of targeting. However, what is the acceptable threshold to be counted as 'effective' is a matter of political judgement. Recent debate in the journal *Environment and Planning A* (Dorling (ed.), 2001) shows that the verdict over people-based and area-based policies remains inconclusive. It is, nevertheless, interesting to see that the new 2004 Index of Multiple Deprivation has categorically detached itself from such a sensitive debate.

> The presentation of data at different geographical levels should not be taken to imply assumptions about the fundamental causes of deprivation, nor should it be taken to imply assumptions about the appropriate solutions. The identification of deprived areas may be necessary if area-based solutions to deprivation are to be pursued, but identifying deprived areas in no way assumes that such solutions are the right ones.
>
> (Noble et al. 2004: 13)

VULNERABLE GROUPS

Indicators such as lone-parent families and those with low socio-economic status are often included in deprivation indices. This is because these vulnerable groups are found to be a good predictor of deprivation (Berthoud 1983). The use of a predictive modelling approach is acceptable, as it is based on a number of variables to predict the level of deprivation, which will lead to the development of a weighted deprivation index. For example, Gordon and Forrest (1995) derived the weighting scheme for the Breadline Britain Index by regressing a subset of indicators (which corresponds to those from the 1991 Census) with the number of deprived households from the Breadline Britain survey data. The weighting was then applied to the selected Census variables to estimate the number of deprived households. This approach is, nevertheless, difficult to implement because most indices are not underpinned by any statistically based modelling procedures. Without the modelling procedures, the inclusion of vulnerable groups in the index can easily run the risk of double counting, as they are proxies of other direct measures of deprivation. This leads to a debate over whether vulnerable groups or groups at risk of disadvantage, should be used as proxy measures.

Townsend provided a very useful conceptual distinction on such matters. He argued that, 'even if many such people are deprived, it is their deprivation, not their status which has to be measured' (Townsend 1987: 135). In reviewing the 1981 DoE Deprivation Index, Coombes et al. (1995) made it clear that, as far as possible, deprivation indicators should aim to measure the event of deprivation directly rather than the 'at risk' groups. This partly aims to avoid the problem of potential double counting, but is partly related to the fact that when the circumstances of a particular social group change, it will no longer serve as a good predictor any more. For instance, following the trend of rising divorce rates, there are now more single-parent families that belong to the more affluent social groups and hence they will no longer serve as a reliable proxy measure of income deprivation. More importantly, it will be politically sensitive to label a particular social group as deprived in an official deprivation index. This line of argument was accepted by the Oxford research team in their review of the Index of Deprivation (Noble et al. 1999: 3), and they shifted the DoE's previous practice from measuring 'surrogates for deprivation' of groups 'at risk' to more direct measures of deprivation. This principle of including direct measures of deprivation has continued in the development of the 2000 and 2004 Indices of Multiple Deprivation.

SPATIAL SCALES AND ADMINISTRATIVE AREAS

Some commentators argue that the crux of the debate over area-based measures and area-based policies lies on the delimitation of the administrative areas, which is dependent on the choice of geographical scales and the selection of their boundaries (Johnston et al. 2003). Geographical scales are important because the variability across different administrative areas tends to decrease at larger spatial scales (e.g. at local authority district level), whereas intra-area variability tends to increase within larger areas. This means that the selection of boundaries can determine the demographic mix and hence an area's homogeneity. Openshaw (1984) called this the 'modifiable areal unit problem'.

The tension thus lies with the mismatch between the administrative geographies and the appropriate functional spatial scale in measuring a particular type of deprivation. When the spatial unit used is too large (i.e. overbounding), hetero-geneity of demographic mix will be introduced to the measurement (Robson *et al.* 1995). The spatial unit is deemed to be too large to capture situations such as poor housing conditions that may be concentrated in pockets of deprivation smaller than the size of the unit. On the contrary, the units may be too small (i.e. underbounding) to record statistically significant incidence of issue such as babies with low birth weight. The problem caused by underbounding of the spatial unit is, however, seen as less problematic (Kearns *et al.* 2000).

The reliance on a single administrative geography in compiling indicators has a range of potential problems, particularly so, if the intention is to track deprivation over time (see Robson *et al.* 2003). The unit chosen may cease to be used for the collection of some input data due to a revision of boundaries as frequently found in electoral ward boundaries. A similar problem is found with the widely used post-code geographies, as they are administrative units of the Post Office. Boundaries devised for administrative purposes will be revised purely on political or business grounds. Any such boundary revisions tend to cause tremendous problems when converting the spatial units back to a consistent set of geographies for time series analysis. For instance, any conversion from postcodes to wards or local authority districts will normally have a degree of error of about 1–2 per cent (Robson *et al.* 2003).

WEIGHTINGS, STATISTICAL STANDARDISATION AND

TRANSFORMATION

As discussed in Chapter 6, different approaches can be used to derive a weighting scheme to combine indicators into a composite index. Even following the same methodology, Saunders (1998) found that unemployment had more influ-ence on deprivation in the Greenwich sample than in the national one used by Gordon (1995) and that the two sets of weighting produced different estimates of the number of deprived with a differential of 1,250. One major criticism of the 2000 Index of Multiple Deprivation centres on the way the weightings were derived for its constituent domains (Chalmers 2001; Deas *et al.* 2003). The precise basis for adopting the weighting scheme for the six domains has never been clarified, beyond the assertion that 'the weightings to combine domain scores were not to be the product of a statistical exercise but included a value judgement' (Noble *et al.* 2001: 7). Elsewhere, the documentation accompanying the index's preparation suggests that: 'the domains with the most robust indica-tors should be given the greatest weight' (Noble *et al.* 1999: 26). This means that the weighting scheme is neither underpinned by a theoretical framework nor based on any empirical model. Yet this weighting scheme, as demonstrated by Chalmers (2001), is critical in the derivation of the index values, and has important policy implications as resource allocation decisions are based almost entirely on composite values rather than domain scores. Following Chalmers, Tomlinson and Kelly (2003) conducted a similar exercise to look at the impact of varying the weightings used in combining the domains of the Northern Ireland's 2001 Index of Multiple Deprivation (Noble et al. 2001). They found that the analysis produced significantly different results (up to 10 per cent or so of deprived wards that might be defined in or out of the worst 10 per cent band) depending on the weightings used. They thus questioned the research team's assertion that due to the high correlation between different domains of deprivation, different domain weightings produced very similar overall rankings of wards.

Besides weightings, other technical procedures such as statistical standardisation and transformation are equally controversial. There are different ways to express indicator values. There is no doubt that expressing the value in percentage is more accessible to policy-makers than other techniques such as z-score and chi-square value. However, the problem lies with finding a relevant population denominator (such as economically active population rather than the entire population) to express a particular indicator value in percentage term (Simpson 1996). Hence, the 2000 Index of Multiple Deprivation only provides percentage value for its income and employment domains. Of the various standardisation methods used in producing the official deprivation index, the use of chi-square in the 1991 Index of Local Conditions (Robson et al. 1995) has been most controversial. Due to the barnardisation of Census enumeration district data (both the 1981 and 1991 Censuses added or subtracted one to the figures on a quasi-random basis to protect confidentiality), chi-squares were used by Rhind (1983) as a compromise solution to deal with the greater unreliability of percentages in small areas. However, when applying the technique to the deprivation index, it took account of both the absolute and the relative size of the deprivation indicator when compared with its expected (national average) value. By doing so, it was effectively dealing with absolute size of an area and the intensity of deprivation at the same time (Connolly and Chisholm 1999). Consequently, this created a scale dependency effect by putting smaller areas in a disadvantageous position. The signed chi-square method was modified in the 1998 revised index (DETR 1998b) by setting the negative values to zeros to avoid the cancelling effect among indicators, but Connolly and Chisholm (1999) proved that the changes actually sharply increased the sensitivity of the aggregate score to the population total of an authority. The index thus focused on the higher-scoring areas (Kearns et al. 2000). Hence, it caused significant distress in smaller authorities, especially the rural areas (Countryside Agency 2000).

The statistical transformation used to dampen extreme distribution of indicator values is another area that causes controversy (see Connolly and Chisholm 1999; Deas *et al.* 2003; Longford 2001). The shrinkage procedures used in the 2000 Index of Multiple Deprivation were seriously criticised by Longford who argued that, 'factor analysis of shrinkage estimates is a terrible practice' that created a false sense of scientific certainty (Longford 2001: 2–3). This echoes Martin *et al.*'s concerns (1994) that transformation tends to lose information that may be of importance to policy and add an extra layer of technical barrier to potential users' understanding of the index. Hence, the use of any transformation procedures has to be well justified.

MEASUREMENT ISSUES: DO THEY MATTER?

The above discussion shows that the derivation of area-based indicators is vulnerable to a variety of conceptual and methodological pitfalls. The question is whether the debates over all these technical issues really matter. A number of review studies find that, irrespective of the methods used and the variables included, striking similarities among different deprivation indices are found (Gordon 1995; Kearns *et al.* 2000). In examining six different deprivation indices for Scotland, Kearns *et al.* (2000) found that they were all highly correlated, with the coefficient ranging from 0.63 to 0.84. To a large extent, high correlations are expected for large-scale spatial analyses, especially when they are conducted at the broad district level. Also, the correlation coefficients of 0.63 to 0.84 only suggest a less than 40 to 65 per cent overlap. Such broad patterns of similarities, however, tend to obscure variations in the rankings between different indices (Gordon 1995; Lee *et al.* 1995). As Gordon found, the specific rankings of areas varied considerably, when looked at more closely, and that the difference became more marked when similar and more homogeneous spatial areas were compared. This suggests that methods and indicators used in constructing the index do matter, and that they can become the target of intense political debate if resources are then attached to the index value.

Throughout the years, the methods used to develop deprivation indices have become more sophisticated and complex. There is a preference toward developing the index at different spatial scales to form a hierarchical nesting structure and embracing a wide range of measures into a composite index (Connolly and Chisholm 1999). The latest development of the Office of the Deputy Prime Minister's (ODPM) 2000 and 2004 Indices of Multiple Deprivation (Noble *et al.* 2000b, 2004) are the case in point. This partly symbolises the maturity of developing deprivation indices after two decade's experience and expertise. It is, nevertheless, a response to pressing political demands of having a deprivation index that can address a whole array of issues and circumstances raised by different stakeholders and researchers. The art of creating a sophisticated index also concomitantly reduces the transparency of the methods and approaches used, which in turn makes it less possible for other researchers to obtain sufficient information to scrutinise the index. The interaction between politics and technical development of deprivation indices will be further examined in the final section of this chapter.

DEPRIVATION INDICES AND THEIR POLITICAL MEANINGS

POLICY TOWARDS PLACES

After examining the different types of policy indicators, Wong (2000) concluded that deprivation indices were more conceptually developed and embedded to the decision-making process than the others. In spite of the flaws and confusion over their policy usage in bidding for the Single Regeneration Budget monies, deprivation indicators were found to be more institutionalised than either sustainability or regional indicators. This is partly due to their long history of development, and partly related to the specific policy needs of making use of such information for resource allocation. Hence, there has been continuous effort to refine and update the indicators. It is the intersection between politics and technical procedures that poses continuous challenges to both policy-makers and researchers. The resource allocation aspect of an area-based needs index is controversial because of its implications for social equity and social justice. First of all, any particular selection of indicators will highlight the problems of one area whilst playing down the problems of another, which leads to different policy outcomes (Fieldhouse and Tye 1996). Second, there are strong reservations over the effectiveness of using area-based approach to eradicate the deprivation problems of individuals. Since all area-based deprivation indices have some sorts of methodological weaknesses, the attention is thus focused on the ways these indices should be used and whether area-based policy should be continued. With regard to the first problem, Tomlinson and Kelly (2003: 83) argued that, due to the sensitivity of deprivation indices to different sets of weightings, it was easier to support the use of deprivation indices for identification, monitoring and evaluation than to support them as a basis for the distribution of resources. The irony is that the main purpose of spending a lot of resources on developing these indices is to serve the hard and fast function of formulae-based resource allocation. This then shifts the nature of the debate from a technical to a political one of whether area-based policy is effective.

In examining unemployment problems, Gordon (1999) and Green (2001) both agreed that the concentration of unemployed had been shaped and sustained by a number of processes operating within the labour market, as well as the interaction of housing-market processes, which led to concentrations of low-income, benefit-dependent households in certain rented social housing estates. As a spatial planner, I would concur with their view that the central argument of having area-based policy is to broaden the opportunity within the wider market to improve the demand as well as the supply side of the economy and its interaction with other social processes, which may in turn improve the opportunity for the individuals concerned. The view from several commentators is that area-based policy should be effective in targeting the deprived individuals. However, if we move away from the micro-perspective, one can argue that area measure should be interpreted as a measure of risk. As illustrated in Coombes et al's study (1995), to say that a particular area has a 40 per cent unemployment rate does not mean that every one within the area is suffering from 40 per cent unemployment. In fact, only 40 per cent of individuals are suffering at the 100 per cent level. This interpretation is consistent with Townsend's (1987) definition that the focus of deprivation is on conditions, should they be the physical, environmental and social states or circumstances rather than purely resource based as poverty is defined. It is under this interpretation that area-based policy makes sense, as resources are injected into neighbourhoods that are at risk of deprivation (Lupton and Power 2002). The problem of poor environmental quality, transport and housing issues are found to be more effectively tackled at the neighbourhood scale. More importantly, there is a general consensus that area-based initiatives cannot substitute, but rather complement, for policies targeted at individuals. This is best summed up by Joshi who asserts that, 'Policies towards Places are not redundant, but they should operate within a context of Policies towards People' (Joshi 2001: 1352).

INDEX REVIEW AND DEVELOPMENT

Since deprivation indices are deeply embedded in the policy-making and resource allocation processes in the UK, it is interesting to explore the review process of such indices and how successive indices were produced. In order to have a more balanced view to gauge the key issues involved, the recent review of the US official measure of poverty will shed some light on the debate. The US official measure of poverty was originally developed in the early 1960s as an indicator of the number and proportion of people with inadequate family resources (before-tax money income) for needed consumption of food and all other goods and services (Betson et al. 2000: 87). Hence, it is a threshold measure to be updated annually to take into account the change in the Consumer Price Index. In response to the growing concerns of the validity and usefulness of the official US poverty measure in the 1980s, the Congress set up an independent panel to review the measure. The Panel on Poverty and Family Assistance of the US Committee on National Statistics was then charged to review and to develop an acceptable and feasible poverty measure in 1992–4. The recommendation of the panel was based on the best scientific evidence available, its best judgement, and three additional criteria:

> A poverty measure should be acceptable and understandable to the public;
> Should be statistically defensible;
> Should be feasible to implement with data that are available.
>
> (Betson et al. 2000: 90)

The panel published its report in 1995 (Citro and Michael 1995) and proposed a new poverty measure to provide a more accurate picture of poverty in the USA and, more importantly, the new measure should allow the tracking of changes of poverty over time that result from new government policies and socio-economic change. It is interesting to note that the panel paid special attention to the importance of transitional arrangements (Betson *et al.* 2000). It proposed that a concurrent poverty measure series, one for the existing measure with its thresholds updated for price changes only and another series for the newly proposed measure, should be published for a certain period of time. Finally, the panel recommended the Statistical Policy Office in the US Office of Management and Budget to develop a mechanism to allow regular review of the poverty measure in terms of its concept, methods and data on a 10-yearly basis. Since the publication of the review, a considerable amount of research effort has been made to flesh out the key issues and to lay the groundwork to implement its recommendations (see Institute for Research on Poverty 1998).

In the UK, since the development of the 1981 Deprivation Index, the DoE and its successors have been playing a key role in leading the development of successive deprivation indices. However, there has been a lack of a long-term strategy over the review and the development of these deprivation indices. The review tends to coincide with the time when new Population Census data becomes available because most small-area based statistics are Census dependent. The review and the development of the deprivation index tend to be included as one of the research projects to be contracted by the Office for the Deputy Prime Minister and its predecessors in its annual research programme. Consultancy firms and researchers can then put in an expression of interest for the project and a selected number will be invited to submit a full research tender and to be interviewed by a selection panel. Since it is through a competitive tendering process, even though the quality of research is the foremost important criterion, other factors such as value of money do come into play in the selection process. While in the USA the review is conducted by an independent panel of experts, in England very often the index has been reviewed and then developed by the same consultant. This casts doubt on the objectivity of the review and may potentially fail to optimise the wider spectrum of expertise from the statistical and academic community. The mitigating measure is to carry out a major consultation exercise. With the advent of Internet technology, web-based dissemination and a major consultation exercise were very much the key features of the development process of the 2000 index.

Another major concern is that whenever a deprivation index was developed in the past for England, the same research team and similar approaches and methods were then employed to duplicate the index for Scotland, Wales and Northern Ireland. This was found especially problematic when the chi-square method of the 1991 Index of Local Conditions was used for the rest of the UK, as these areas are largely made up of smaller and more rural areas. As discussed earlier, these areas tend to fair badly under the chi-square method. Such an English-led approach towards deprivation indicators development continued after the publication of the 2000 Index of Multiple Deprivation; the same research team was employed to review and develop respective indices for other parts of the UK. It is thus not a surprise to see that the same methodology and approaches were used in these indices, although with minor adjustments to the number and definition of indicators. This steamroller also extended to the territory of the Countryside Agency. Despite its own commissioned research in developing a new approach to rural disadvantage, undertaken by Cambridge University (Dunn et al. 1998), it still jumped on the bandwagon when the 2000 English index was published. This led to the publication of a consultation report of indicators of rural disadvantage for each English region in which the proposed indicators were drawn from the 2000 Index of Multiple Deprivation; its own commissioned indicators of rural disadvantage and information related to the Rural Services Survey (Countryside Agency 2001). One of the main reasons behind such a convergent approach towards the development of deprivation indices is probably driven by the need of comparability, even though at times the methodology may not be suitable for measuring the circumstances and policy needs of specific localities.

The third concern is the lack of an established mechanism to oversee the review and development of deprivation indices, though they have been deeply embedded in the policy-making process. Unlike the US panel, there is a lack of commitment in developing indices that allow change analysis over time. The volatility of the methods and indicators used in the deprivation index means that it is not easy to carry out trend analysis. This problem is epitomised in the press release of the London Borough of Haringey,

> The ID2004 found that Haringey is the 10th most deprived district in England, as measured by both the average of ward ranks and the extent of deprivation. When compared to the previous ID2000 this represents a much more severe scale of deprivation, although this is most likely to be due to changes in the way the 2004 index was calculated and the data used than to a real relative increase in deprivation.
>
> (Haringey Council 2004)

The lack of consistency in the use of a set of widely accepted indicators to gauge change over time has raised concerns by academics, as it prevents us from knowing about the process of neighbourhood and urban change (e.g. Lupton and Power 2004; Wong *et al.* 2004). One of the major achievements of the 2000 and 2004 Indices of Multiple Deprivation has been the use of administrative and survey data sources to develop indicators for small-area analysis. However, as the research team themselves point out, survey estimates are more useful in providing a cross-sectional snapshot of the relative performance of neighbourhoods, but not robust enough to examine absolute change to inform trend analysis as they may just measure change in standard errors rather than genuine changes (SDRC 2004). The use of administrative data also means that it is not that straightforward to carry out longitudinal analysis when there are changes in policy/benefit claiming regulations (Tomlinson and Kelly 2003). Since there is little scope for making meaningful comparison of successive deprivation indices, any change in the rankings as a result of the release of a new index is simply showing a change in the relative position of an area in comparison with others rather than the actual change in an area's deprivation conditions. This change can, nevertheless, bear significant resource implications to local authorities. Hence, it makes sense to consider transitional arrangements, and the case of making concurrent publication of the old and new indices for a short period of time, along the line recommended by the US poverty review panel.

In the latest round of deprivation index development, Scotland has taken a somewhat different approach. The Scottish Neighbourhood Statistics set up an Index of Deprivation Working Group in 2001 with the remit to develop a Scottish Index of Deprivation that would be used for resource allocation to and within local authorities (SNS 2004). The Working Group took the task forward by splitting the work into two strands: first to compile an interim index based on the same methodology used in the 2000 English index, and second to develop a long-term strategy for index development. Members of the

Working Group included different divisions of the Scottish Executive, local authorities, Scottish Homes, Communities Scotland, health boards and National Health Service Scotland. The Group held five meetings between October 2001 and December 2003, and was disbanded after it successfully fulfilled its remit at the end of 2003.

The Social Disadvantage Research Centre at Oxford University, the same research team who previously developed the English/Welsh/Northern Ireland indices, was commissioned to develop the 2003 Interim Index. The Scottish Centre for Research on Social Justice was asked to develop a long-term strategy for measuring deprivation in December 2002. Through a steering group, consultation on the interim report and public meetings, the views of central and local government, community groups, academics and the wider general public were sought. The final report was published in September 2003 and the Scottish Executive accepted all its recommendations. There are several major areas of development in respect of the longer-term strategy. First of all, the 2004 Index of Deprivation is due for publication in spring 2005, which is a revised version of the 2003 Interim Index with updated information and additional datasets (SNS 2004). The 2004 Index is heavily based on the methodology developed by Oxford University but, in contrast to previous prac- tice, 'the Scottish Executive has taken the responsibility for sourcing the data, constructing the domains and quality assuring the projects' (SNS 2004). This contrasts with the situation in England, as the Index of Deprivation is not part of the National Statistics. The Scottish Executive makes a commitment of ensuring the quality of the index by taking responsibility for its methodological advancement and the use of new data sources. More importantly, a clear timetable is set out, including a commitment to update the 2004 area-based index in October 2006 and thereafter on a three-year cycle. Finally, one of the important recommen- dations from the long-term strategy report is that there is a need to provide and maintain an area-based approach and an individual approach (Bailey et al. 2003) to allow policy users to apply the measure that is most relevant to their needs. The proposal of allocating resources to conduct large-scale social survey such as the Poverty and Social Exclusion Survey, and making use of administrative records held by government on individuals, is widely welcome as it will allow the development of direct measures of deprivation for individuals and different social groups.

It is clear that the Scottish Executive has not gone as far as the USA in appointing an independent expert panel to make a wholesale review of the measures. It retained the traditional approach of commissioning consultants to do the review that is overseen by a Steering Group (in this case, the Index of Deprivation Working Group), alongside a major consultation exercise. However, it does demonstrate that a longer-term strategy and a timetable are now in place to take the process of index development forward. This commit-ment is further reinforced by the fact that the Scottish Executive is willing to take full responsibility over quality assurance matters of future indices and has made it clear that the index will be used for resource allocation purposes. The Scottish approach thus moves a big step in the right direction towards estab-lishing a framework for longer-term development and to inject a sense of stability over the relationship between the technical methodological develop-ment of deprivation indices and the political usage of resource allocation.

CRITICAL FRIENDS

Since the 1981 Deprivation Index, the development of successive indices has caused significant attention from the policy and academic community. As discussed in Chapter 3, such political attention is related to its subsequent use as a resource allocation tool. Hence, local authorities that fare less well from these indices would be keen to make challenges on methodological grounds. Reputable experts and academics in the field were thus commis-sioned to examine whether there were any technical flaws in the index to merit a challenge. This created a very difficult situation as those who were commis-sioned by a particular organisation to do the review could be accused of being biased and their criticism could thus be easily dismissed. After the publi-cation of the 1991 Index of Local Conditions, and its revised 1998 index, Professor Michael Chisholm, an eminent economic geographer from Cambridge University, had been working closely with Durham County Council to make a critique of the methodology used in the index (e.g. Connolly and Chisholm 1999). However, the debate became more heated after the 2000 Index of Multiple Deprivation was released. A discussion forum on the new 2000 index was held by the Royal Statistical Society on 11 July 2001. The speakers that contributed to the event included two academics from the London School of Economics, Jane Galbraith and Colin Chalmers, and Nick Longford from De Montfort University. However, the event and the criticism made by Chalmers caused significant ripples. The research team at Oxford University wrote an open correspondence to the *Journal of Royal Statistical Society A* to express their concerns and dismay. In the letter, they made some explicit state-ments:

> For the record, in 2000 Mr. Chalmers was employed as a consultant by the Greater London Authority explicitly to provide a review of the new indices of deprivation. It is a matter of public record that the Greater London Authority has campaigned vigor-ously against the indices, as they are fully entitled to do. . . .

> We are firmly against censorship, but equally we would wish to see the journal's high
> reputation protected from the damage that might be caused by the publication of such
> unrefereed material where there is no right of reply.
>
> (Firth *et al.* 2001: 566)

From this incident, it is clear that there is a certain degree of uneasiness for researchers to engage in applied policy work. This exactly reflects Berger and Kellner's (1982) concern of falling into the trap of dual citizenship (see Chapter 2). The crux of the debate has gone beyond the original concern over technical statistical procedures used in standardising the index. As an observer, and someone who has also engaged in reviewing the development of indices for different bodies, I can see part of the problem is caused by the shift towards using more complex modelling techniques and different sources of confidential data to estimate indicator values. Since the release of raw data for some indicators is restricted by the Data Protection Act, the whole process of index development can no longer be totally transparent to allow close scrutiny. It is the complexity of the index, in combination with the absence of actual indicator values, that makes it difficult to be a 'critical friend' and can easily lead to misunderstanding and mud-slinging episodes. There is a real trade-off between the statistical black box and the desire to obtain estimated small-area statistics for all sorts of indicators. The complexity leads us to believe that a better index is created – this may or may not be the case, as the privileged information is only made available to the commissioned research team. I think it is this lack of transparency that stifles debate and may hinder further progress in developing such indices.

FINAL THOUGHTS

As the policy objectives become more multi-faceted and more demanding, it is important to employ a more flexible, modular approach that suits different policy needs under the evidence based policy regime. Even far back to the 1981 DoE index, there were sub-indices with different weightings, and the *Better Information* report also pointed out that different deprivation measures were used by different government departments for resource allocation. The Scottish strategy of moving towards measuring deprivation for individuals is admirable, though there is a need to have a real commitment of resources. There is an urgent need to develop some mechanisms to govern more independent reviews of indices that will be used to allocate public resources and to devise methods to allow external scrutiny of such indices. (In the light of the discussion above, it is obvious that the use of an independent panel is more credible than using competitive tenders!). To draw an end to this discussion, I would like to borrow a quote from Lee *et al.*, 'this analysis may leave the reader with the conclusion that all the indexes have their limitations and that the pursuit of a best index is, at best, an impossible holy grail' (Lee *et al.* 1995: 54).

CHAPTER 9

URBAN AND REGIONAL DEVELOPMENT
INDICATORS

TRACKING URBAN AND REGIONAL CHANGE

The use of social measures to improve our understanding of the main features of society, how they inter-relate, and how these features and their relationships change was started way back in the 1930s and 1940s by William Ogburn at the University of Chicago (Land 1975; Sheldon and Moore 1968). This produced what Land (2000) called descriptive social indicators of the state of society. The need to have such contextual indicators to profile characteristics of different places is increasingly important in the process of formulating urban and regional development policies. Following in the footsteps of the social indicators movement, supranational organisations across the world have shown their interest in assessing the state of urban development across different nations. The World Bank (2003) compiles the annual world development indicators series to monitor the achievement towards international development goals. Eurostat has redeveloped 'Urban Audits' to improve the European Commission's knowledge of quality of life in urban areas across Europe (European Commission 2000b). A similar venture can be found on the other side of the globe. The Asian Development Bank recently has made concerted effort to compile the Cities Data Book to inform the management of the urban sector in Asia (Westfall and de Villa 2001). In England, the government is committed to monitor the urban renaissance visions set out in the *Urban White Paper* (DETR 2000d) by developing a Town and City Indicators Database (see Wong *et al.* 2004) and publishing the State of the Cities Reports (e.g. Robson *et al.* 2000).

Rather than being the most politicised, indicators on urban and regional development tend to appeal to a wide range of stakeholders and are frequently under the limelight of the media. This is because indicators measuring characteristics of places tend to find their way into place ratings and league tables. The chasm between traditionally highbrow academic research and the popular public and business interests can thus be bridged over by such statistics. However, the perplexing issues involved in the development of these indicators do not always produce what they appear to promise. Quality of life and economic competitiveness indicators are the foremost examples of these media-friendly indicators. The next section will start off the discussion by tracking the ups and downs of the usage and development of place liveability and competitiveness research in the USA and the UK. The politics and flaws of these studies will be highlighted. Attention will then turn to explore the use of indicators in monitoring regional economic policies, and especially on the progress made in developing the intelligence and institutional structure to support the devolution of regional policy in England. Finally, some remarks are made at the end of the discussion.

Quality of Life and Place Competitiveness studies

The fashion of comparing and ranking places based upon a synthetic composite index swept the USA in the 1980s through the publication of the *Places Rated Almanac* (Boyer and Savageau 1981, 1983, 1985). This approach of study focuses on examining the quality of the shared living environment in cities through measuring a number of objective indicators from secondary data sources such as house prices, income, education levels, health care, climate, parks and public space. These urban liveability studies are widely known as quality of life studies (Myers 1988), though Landis and Sawicki (1988) emphasised that such studies really addressed 'quality of place' rather than the quality of life of individuals. The public acclaim for the *Places Rated Almanac* was attributed to their wide publicity (Myers 1988), but also to the fact that they were a serious attempt to popularise a statistical ranking of metropolitan areas (Rogerson 1999).

In spite of enjoying success in capturing the public imagination, the *Places Rated Almanac* was widely criticised for its methodological flaws (Becker *et al.* 1987; Bell 1984; Landis and Sawicki 1988), especially when compared with some earlier quality of life studies (e.g. Smith 1973; Liu 1976). The criticism of recent US liveability studies tends to focus on several areas (Myers 1988: 353): the subjective assumptions made by researchers on the definition of quality of life; the lack of a robust theoretical framework to guide measurements; *ad hoc* use of available data; arbitrary selection of indicators and weighting schemes; and the production of erroneous ratings of quality of life. Landis and Sawicki (1988) made a very detailed evaluation of the *Places Rated Almanac* by examining its conceptual appropriateness, indicator reliability, ecological fallacy, scaling issues and double counting, and policy relevance. They concluded that the *Almanac* might be of some use to footloose migrants, but it contained some fundamental conceptual and measurement weaknesses, and would be of little use to enhance our understanding of the quality of life in localities. The response from the *Places* authors to the critics was simply that, 'With so much in life that can't be predicted, it's necessary to supplement *Places*

Rated Almanac with your own independent verification' (Boyer and Savageau 1985: xiii).

Land (2000) suggests that the widespread political, popular, and theoretical appeal of the quality of life concept is due to its integrative and unifying nature, which encompasses all domains of life, including an individual's material and immaterial well-being and the natural conditions of life for present and future generations. The multi-dimensional nature and the ambiguity of the concept (Myers 1988) means that policy-makers are more than happy to embrace it, and different stakeholders may have their own interpretation of what quality of life may mean (Harwood 1976). Schuessler and Fisher thus argue that quality of life 'functions in a meta-theoretical way to reference research aimed at policy outcomes' (Schuessler and Fisher 1985: 132). There is a consensus among researchers that a precise and universally accepted definition of the concept has yet to be framed and the challenge comes when attempting to measure the latent traits of the concept that are not susceptible to direct observation. The measurement of quality of life issues is technically complex and, if not properly implemented, the findings can easily be subject to misinterpretation and political manipulation.

In defining quality of life, economists tend to use income levels and house prices in their measurement (Rosen 1979; Roback 1982; Stover and Leven 1992), whereas the urban liveability approach tends to define it as purely non-marketable public goods such as climate, environmental amenities, crime, traffic and public services (Liu 1976; Findlay *et al.* 1989; Boyer and Savageau 1985). However, these non-marketable public goods are increasingly seen as vital in fostering economic growth and job creation by retaining local businesses and attracting inward investors (Schmenner 1982; Hall *et al.* 1987; Bosman and de Smidt 1993; Johnson and Rasker 1995). The causal relationship between local economic development and quality of life is, however, a difficult and controversial topic (Wong 2001). Castells (1989: 52) regarded quality of life as a *result* of the characteristics of the high-tech industry in Silicon Valley (its newness and highly educated workforce) rather than the *determinant* of its location pattern. Findlay *et al.* (1989) also failed to find any significant correlation between their quality of life index and the local prosperity index (Green and Champion 1989) in British cities. My own empirical research (Wong 1998a, 2001, 2002a) found that quality of life was more important in shaping the reproduction, rather than the production, space of an area. Quality of life alone can provide a very desirable living environ-ment for commuters who then travel outwards to take up high-quality jobs elsewhere. Nevertheless, good living quality does provide the cutting-edge in the competitive process when a number of potential investment locations are on a level playing field in terms of traditional production factors (Wong, 1998a). Regardless of whether it is the determinant or the outcome of economic devel-opment, quality of life has clearly become an asset in place marketing to lure talents, attract mobile global capital (Rogerson 1999) and influence relocation decisions (Bovaird, 1995). The commodification of quality of life research leads to the danger of simply reducing the concept to those characteristics desired by inward investors and neglects the viewpoint of other stakeholders (Rogerson 1999), which in turn reinforces existing spatial disparities in quality of life (Bovaird 1995).

The interest in producing comparative place ratings of cities also spread to the UK (see Rogerson 1999; Wong 2003) in the late 1980s. The growing need to identify urban regeneration potential and needs, and increasing policy interest in statistical information, helped spur significant academic contributions in measuring socio-economic conditions and prosperity. These included the 'booming towns' study by Champion and Green (1990), Breheny et al.'s (1987) 'northern lights' research, the 'quality of life' study by the Glasgow quality of life research team (Rogerson et al. 1989) and the measurement of 'Superprofile' geodemographics by Brown and Batey (1994). These research studies have rejuvenated academic debate over various methodological issues, for example, different ways of quantifying intangible issues (Rogerson et al. 1989); derivation of appropriate weighting systems to create composite indices (Green and Champion 1989; Rogerson et al. 1987); and the use of innovative methods to develop area classifications (Brown 1989; Brown and Batey 1994). On the whole, rather than bearing any specific policy concerns in mind, these academic studies tended to produce place-rating schemes based on socio-economic performance or geodemographic profile. The outcome of the resulting league tables was often controversial, partly because they were subject to misinterpretation by the media. However, such media hype on city development league tables has raised the attention of policy-makers, as well as furthered academic interest, in quantitative measurement. The nature of this wave of urban indicators research in Britain tended to be strongly grounded in an empirical approach. Most research effort has thus been paid to the methods of measurement and the empirical exploration of data.

In recent years, quality of life studies have shifted from academic circles and popular interest to the mainstream policy agenda. The Audit Commission claimed that it was due to a number of reasons:

> The Local Government Act 2000 (HM Government, 2000) placed a duty on local authorities to produce a long-term community strategy with their partners to improve the quality of life in their local area. At the same time, following the international Earth Summit in Johannesburg in Aug/Sept 2002, there is increasing pressure on local authorities and their partners to ensure that their activities and plans are based on the principles of sustainable development.
>
> (Audit Commission 2002: 2)

It is clear that the international movement of sustainable development and central government's policy agenda have embraced quality of life indicators as part of the administrative requirements. The Audit Commission, therefore, took the lead in a year-long pilot project to develop a set of quality of life indicators with more than ninety local authorities during the financial year 2001/2. The pilot exercise identified thirty-eight indicators under thirteen themes:

combating unemployment,
encouraging economic regeneration,
tackling poverty and social exclusion,
developing people's skills,
improving people's health,
improving housing opportunities,
tackling community safety,
strengthening community involvement,
reducing pollution,
improving management of the environment,
improving the local environment,
improving transport, and
protecting the diversity of the nature.

However, this is not the only set of quality of life indicators in the policy domain. The government has already issued a set of thirteen Quality of Life Headline Indicators in 1998. This was the result of narrowing down from 120 Indicators of Sustainable Development (DoE 1996) as proposed in the *Sustainability Count* consultation document (DETR 1998a) published in November 1998 under the framework of four objectives derived from the revised definition of sustainable development in the *Opportunities for Change* paper (DETR 1998e). However, the reincarnation of sustainable development into quality of life indicators has caused concern: 'whether the resulting package of indicators actually measures sustainability remains unclear, in that it is only loosely pinned to a meaningful concept of sustainability' (Atkinson 1998: 307). The revised definition of sustainability strongly emphasises the 'quality of life' component, and the Deputy Prime Minister called the associated indicators 'quality of life' headline indicators (DETR 1998a: Foreword). As discussed earlier, the multi-dimensional nature of 'quality of life' means that it can easily embrace the meaning of 'sustainable development' – even though the two concepts are analytically distinct, as it is possible for one to occur without the other (Wong 2000). When comparing quality of life and sustainability indicator sets, Swain and Hollar (2003) found that the latter was much more driven by a particular set of visions and values, and relied on an ecological frame of reference to reveal the interconnectedness over time and across space.

Table 9.1 1998 and 2004 Quality of Life Count Headline Indicators

Objective 1: maintenance of high and stable levels of economic growth and employment	
Economic growth	1998: total output of the economy (GDP)
	2004: GDP and GDP per head
Social investment	1998: investment in public assets (transport, hospitals, schools etc.)
	2004: total and social investment relative to GDP
Employment	1998: people of working age who are in work
	2004: proportion of people of working age who are in work

Objective 2: social progress which recognises the needs of everyone	
Poverty	2004: indicators of success in tackling poverty and social exclusion
Health	1998: expected years of healthy life
	2004: expected years of healthy life
Education and training	1998: qualifications at age 19
	2004: qualifications at age 19
Housing quality	1998: homes judged unfit to live in
	2004: households living in non-decent housing
Level of crime	2004: violent crime; vehicles and burglary; and robbery

Objective 3: effective protection of the environment	
Climate change	1998: emissions of greenhouse gases
	2004: emissions of greenhouse gases
Air pollution	1998: days of air pollution
	2004: days when air pollution is moderate or higher
Transport	1998: road traffic
	2004: traffic volume; and traffic intensity
Water quality	1998: rivers of good or fair quality
	2004: chemical and biological river quality
Wildlife	1998: populations of wild birds
	2004: populations of wild birds; woodland birds; farmland birds
Land use	1998: new homes built on previously developed land
	2004: new homes built on previously developed land

Objective 4: prudent use of natural resources	
Waste	1998: waste and waste disposal
	2004: waste arisings and management

Source: DETR, 1999b and DEFRA, 2004.

The statisticians behind the design of the sustainability count indicators actually pointed out that crime was, for example, clearly a quality of life issue but was not directly relevant to sustainable development (Custance and Hillier 1998). It is, however, interesting to find that 'level of crime' was the only indicator added to the original thirteen headline indicators in the following year's progress monitoring report (DETR 1999b).

By 2004, the total number of Quality of Life Headline Indicators had been increased to fifteen and the core set of national indicators had also increased from 120 to 147. When comparing the headline indicators in 1998 with the current set (see Table 9.1), it is clear that indicators on poverty and crime were the two new additions to the baseline set. In addition, some adjustments and refinements were made to the definition of the indicators, especially on indicators measuring economic growth and development, and the environment. It is also interesting to contrast this set of quality of life indicators with those of the Audit Commission. The key themes involved in the two sets of indicators are strikingly similar, as are the indicators. Local authorities are currently compiling two similar sets of quality of life indicators in parallel, one for monitoring the community strategy under the local strategic partnership for the Audit Commission and the other for monitoring spatial strategies to meet with the requirement of sustainable development for the Office of the Deputy Prime Minister and the Department of Environment, Food and Rural Affairs. One has, therefore, to question the rationale for this unnecessary duplication, as it has imposed further burden of information collection on local authorities. It is thus not surprising to hear the suggestion that there is a need for different government departments and public agencies to rationalise and harmonise their monitoring requirements over quality of life indicators. The intertwining of different government indicator sets over the measurement of sustainability will be further discussed in Chapter 10.

Besides quality of life studies, there was also a major debate in the late 1980s over the use of indicators to measure state business climates in the USA. Three organisations, Grant Thornton (an accounting firm), INC (a magazine publisher) and the Corporation for Enterprise Development (a non-profit consulting group and publisher), all generated annual reports to serve such a purpose (Boyle 1989; Skoro 1988). The dramatic difference in the rankings produced in these reports has generated interesting stories for the media, and officials of these organisations were regularly called upon to defend their ratings and attack their competitor's methodology. When comparing the three reports, Boyle (1989) commented that they reached very different conclusions because they asked very different questions. The Grant Thornton ratings were based on twenty-one factors and their relative importance was determined by representatives of manufacturers' associations. The result was thus a 'manufacturing climate index' for each state, and was widely seen as the result of a poll of lobbyists for manufacturers. INC's study measured results rather than location factors. Its rating did not evaluate business climate, nor produce reliable predictors of future performance. The Corporation for Enterprise Development model (CFED, 1991) had four sub-indices: economic performance, business vitality, business capacity, and state government policies. It tried to measure performance, and traditional location factors, as well as community conditions and equity issues. Skoro (1988) commented that the State Development Report Card produced by this model was simply political statements with little verifiable content and could be seen as a reply to the poll of Grant Thornton. The analysis of these business climate indexes raises a number of issues. The foremost question is whether it is wise to measure business climate, as business is not a monolithic entity (Boyle 1989) and different companies within the same industry may value different attributes of the facility. Another main concern is the technical competence employed in constructing these indices; the methodologies used were subject to value bias and the outcomes were used for attracting media attention and making political statements (Skoro 1988). For instance, INC never revealed the weighting scheme assigned to the individual factors.

On this side of the Atlantic, there is less competition in producing business climate indices. The Confederation of British Industry has routinely carried out its regional survey of British companies, the findings of which tend to produce a ranking of corporate priorities for business competitiveness (e.g. CBI 1997). There have been a number of academic studies in gauging the enterprise potential and urban growth in the UK and in Europe. Examples included the development of a 'local enterprise activity potential index' by Coombes and Raybould (1989), the measure of urban growth/decline of 103 metropolitan regions in the EU by Cheshire *et al.* (1986), and more recently the measure of determining factors of local economic development for English local authorities by Wong (2001, 2002c). These academic studies tend to be driven by rigorous methodology and analytical frameworks, and do not set off to invite media attention over the rankings produced. However, under the influence of neo-liberal ideology and the Thatcherite government policy, a range of studies were commissioned by different government departments to identify indicators to measure their specific policy concerns. One interesting example was a project commissioned by the Department of Employment to develop a 'Local Environment Index' modelled on the Corporation for Enterprise Development's State Report Card (see Coombes *et al.* 1993b). Most of these official studies were commissioned to provide policy inputs to a particular government department and tended to be short term and *ad hoc* in nature. Hence, the scope for fundamental theoretical and methodological analysis was highly constrained. For instance, the conceptualisation of key issues such as 'competitive advantage' and 'economic performance' in a report prepared by

PA Cambridge Economic Consultants (1990) appeared to be somewhat inadequate and superficial. Pieda's (1995) practical guidance on local economic audits focused on the steps and procedures of measurement, but key issues and factors included in the proposed framework were not thoroughly explained. In addition, it is worth noting that some of this commissioned work was in the form of pilot studies and did not involve the compilation of a complete data set (Coombes *et al.* 1992; LGMB 1995).

The discussion above shows that indicators on the liveability and competitiveness of places have been around for years and have their ups and downs. Compared with deprivation indicators, they are less well developed in terms of their conceptual definitions and operational methodologies. As argued elsewhere in Wong (2000), indicators that are more embedded into the decision-making process tend to be more conceptually developed and technically sound. As long as the usage of quality of life and competitiveness indicators is *ad hoc*, and without any specific linkage to a particular policy programme, it will be more difficult to see gradual and systematic development to deal with some of the issues identified. Unlike the USA, a lot of indicator research studies in Britain have been commissioned or developed under central government guidance. It is, therefore, interesting to see that, after several decades' work on quality of life research, it is the politicians who have brought such indicators into the heart of the public policy arena. In the pioneering days, Liu already commented that 'the search for quality of life indicators is an attempt to obtain new information that will be useful to evaluate the past, guide the action of the present and plan for the future' (Liu 1976: 3). The concern is whether the current evidence-based policy-making culture in Britain can sustain and embed such indicators into the policy-making process, in order to make a difference on the living quality and competitiveness of places.

DEVOLUTION OF REGIONAL ECONOMIC POLICIES AND INSTITUTIONAL CHANGES

The devolution of regional economic planning in England has gathered paces after the publication of the *Building Partnerships for Prosperity* White Paper (DETR 1997c). Regional Development Agencies (RDAs) are now taking charge of the preparation of Regional Economic Strategies and the allocation of regeneration resources. However, the decentralisation of policy-making power to regional stakeholders has been accompanied by a stringent auditing culture of performance monitoring and evaluation from Whitehall. It is, therefore, interesting to see in what ways indicators have been deployed in monitoring regional policies; whether there are adequate resources to support an evidence-based policy-making regime; and what institutional changes have been made to support the devolution of economic policies to the regions.

THE AMBIGUITY OF REGIONAL COMPETITIVENESS AND STATE OF THE REGION INDICATORS

In the *Building Partnerships for Prosperity* White Paper, the government emphasised the importance of having 'Regional Competitiveness' indicators to track down the performance of each region in areas such as skills, business activity, employment, infrastructure and transport (DETR 1997c: para. 2.5). It made specific reference to the thirteen regional competitiveness indicators in a consultation document of the Department of Trade and Industry (DTI 1997). The first edition of these Regional Competitiveness indicators was subsequently published in February 1998 (DTI 1998) and an updated version has since been published on an annual basis. Meanwhile, a guidance note was issued by the Department of the Environment, Transport and the Regions (DETR 1999a) on the preparation of Regional Economic Strategies. A checklist of core indicators for RDAs was drawn up in Annex II of the document. It consists of ten 'State of the Region' indicators (see Table 9.2) and five specific 'RDA Activity' indicators. When comparing the Regional Competitiveness indicators with the State of the Region indicators (see Table 9.2), one can easily find that there is a high degree of overlap between them. An unavoidable question to ask is the logic of having two sets of indicators, both related to regional development, devised by two separate government departments in a very short span of time. In the DETR guidance, it did mention that in order to improve the quality and range of information for evaluation and monitoring, 'this work will need to take account of existing indicators such as Regional Competitiveness indicators, Quality of Life indicators, Regional Trends and the Index of Local Deprivation, amongst others. It will provide an opportunity to improve the core indicators' (DETR 1999a: 9). However, it is still not clear what exact roles these indicators are supposed to play. Later on, an explanatory note on the DTI website provided further clarification of the role of the 'State of the Region' indicators:

> This [evaluation and monitoring] framework includes a set of the 'State of the Region' indicators which reflect the purposes for which the RDAs were set up. RDAs should use these indicators to inform the development of their strategies. They are high-level regional indicators; they will not be used to judge the performance of individual RDAs, over the first year of their operation at least, although RDAs will want to report on progress against these key regional indicators.
>
> (DTI website 1999)

Table 9.2 Regional Competitiveness and State of the Region indicators

1997 Regional Competiveness Indicators	1999 State of the Region Indicators	2005 Regional Competitiveness and State of the Region Indicators
1. Overall competitiveness • GDP and household disposable income per head • labour productivity in manufacturing • social security benefit claimants • manufacturing investment and output by foreign-owned companies	1. Economic development • GDP per head • GDP per head relative to EU average 2. Competitiveness • Manufacturing and services gross value added	1. Overall competitiveness • gross value added and household disposal income • labour productivity • manufacturing investment and output by the UK and foreign-owned companies • export of goods
2. The labour market • average earnings • employment • unemployment	3. Employment • ILO unemployment rate	2. Labour market • average earnings • employment • unemployment • claimant court • education and vocational attainment
3. Education and training • educational and vocational attainment • Investment in People	4. Skills • % 19 year old with Level 2 qualifications • % adult with Level 3 qualifications • % employers with hard to fill vacancies • % employees undertaking work-related training in last 13 weeks	3. Deprivation • income support claimants • income deprivation
4. Capital • VAT registrations and survival rates • R&D intensity and employees in hi-tech industry	5. Business support • business formation and survival rates	4. Business development • business registration and survival rates • total entrepreneurship activity • R&D, and employment in high and medium-high technology industries
5. Land and infrastructure • transport • industrial property and office rental costs	6. Sustainable development • % new homes built on previously developed land	5. Land and infrastructure • transport • industrial property and office rental costs • re-use of vacant and derelict land

Source: DTI, 1998, 2004; DETR, 1999a.

When carrying out a survey of the monitoring and evaluation frameworks of the Economic Strategies across different regions in 2000 (see Table 9.3), Baker and Wong (2001) found that nearly all RDAs had gone for an all-embracing approach of having different sets of indicators. The most commonly mentioned indicator sets were the official recommended ones from central government – the State of the Region indicators and the DTI's Regional Competitiveness indicators. In addition to these two sets, different RDAs have come up with additional indicators under different labels and branding. These additional indicators included headline performance indicators, region-specific indicators, policy-related targets and performance measures. However, an interesting observation was that none of these Economic Strategies specifically mentioned the headline quality of life indicators in spite of the fact that they were mentioned in the DETR guidance. This suggested that there was an implicit division of ownership of indicator sets, as the quality of life indicators were widely used for monitoring Regional Planning Guidance.

Following a full review of Regional Competitiveness indicators in 2001/2, an interesting move was made by the DTI. In autumn 2002, an amalgamated version of the Regional Competitiveness and State of the Region indicators was published. A number of indicators were excluded from the combined set following the recommendation of the consultants SQW Ltd (DTI 2004). There are now seventeen core indicators in the combined Regional Competitiveness and State of the Region indicators set (DTI 2005). The consultants also identified eleven core indicators out of this combined set for RDA evaluation and performance monitoring (DTI 2002a). As shown in Table 9.2, the combined set mirrors closely the Regional Competitiveness indicators, with some additions from the State of the Region indicators and two extra indicators on income deprivation. The inclusion of a deprivation dimension probably reflects the fact that social regeneration now falls under the remit of the RDAs. The reason for such a rationalisation was explained in a vague and tautological statement, 'They are intended to give a balanced picture of all the statistical information relevant to regional competitiveness and the state of the regions' (DTI 2004: 3). There was also a brief introduction to the nature of the two sets of indicator:

> The aim of the Regional Competitiveness indicators was to present statistical information that illustrated the factors that contributed to regional competitiveness. They were not intended to measure the performance of the Government Offices or the devolved administrations, but were designed to assist those responsible for developing regional strategies. The State of the Region core indicators were originally designed to measure progress towards sustainable economic development, skills and social regeneration and to provide monitoring and evaluation guidance for the RDAs.
>
> (DTI 2004: 3)

Table 9.3 Monitoring and evaluation arrangements of Regional Economic Strategies

Region	Regional Economic Strategy – 2000	Regional Economic Strategy – 2003
South East	• State of region indicators (12) • Region-specific indicators (to be developed with partners) • Monitoring and evaluation framework (to be developed with partners)	• Collaboration with SEEDA and other regional partners working via arrangements such as SE of England Intelligence Network (SEE-IN) and Skills Insight • RES indicators prepared within context of RSDF (prepared by RA, GOSE, EA, NHS & SEEDA). Provides an overall framework, containing 41 indicators • Regional outcome targets and indicators (11), agreed with SEEDA, other regional partners and government, focused on economic priorities • Feed into RES indicators (25) embedding economic priorities within sustainable development
London	• State of region indicators (12) • Regional performance indicators (5) • London performance indicators (based on the 7 recommended in the Association of London Government's London Study) • Possible additional indicators (7)	• London development Agency long-term 'Tier 1' strategic objectives: Economic growth Knowledge and learning Diversity, inclusion and renewal Sustainable development • Medium-term 'Tier 2' regional outcomes (14 indicators relating to the regional economy) • Annual 'Tier 3' programme targets
South West	• State of region indicators (12) • Regional competitiveness indicators (13) • Region-specific indicators (28) • Implementation indicators (to be developed with partners) • Annual monitoring • Possible Regional Observatory	• Eleven Tier 2 targets set by government, supplemented by region-specific targets

South West		• Measurement of progress against these targets to be undertaken at three year intervals, with interim assessments annually
		• Regional Observatories expected to have increasing role in carrying out this monitoring role on behalf of regional partners
Eastern/East Anglia	• State of region indicators (12) • Region-specific indicators on quality of life, performance measures and targets (with regional partners) • Regional Observatory to be established to provide new information and intelligence network • Regional Intelligence Manager appointed to develop the concept of the Observatory	• Three regional performance indicators to measure progress in becoming one of Europe's 'top 20' regions: *Economic output (2); Improved wealth for individuals (1)* • Indicators for 6 major themes: Competitive businesses/organisations (15); Creativity, innovation and enterprise (3); Investing in success (2); Regeneration supporting people and communities (11); Identity and international profile (3); Infrastructure and environment (10) • Additional Quality of Life 'basket' of indicators (9) • Some of these indicators taken from DETR 'Quality of Life Counts' reflecting a desire for sustainable economic development • Some specifically chosen to reflect sub-regional or demographic level • Regional Observatory will make it easier to use broader range of data and information in the future
North West	• State of region indicators (12) • Activity indicators in relation to its programmes	• Indicators to measure performance of the strategy fall into three categories:

Table 9.3 (continued)

North West	• 3–6 year targets linked to its objectives • Region-specific indicators (to be developed based on suitability and availability of data)	• 'Tier 2' indicators for which regional outcome targets are agreed with government; Additional indicators, as far as possible to be consistent with measures included in other regional strategy documents; Quality of life/sustainable development indicators monitored through 'Action for Sustainability' (RSF) indicators • Working with NWRA, GONW and other regional partners to develop harmonised set of indicators to measure overall progress against 'vital signs' of economic, social and environmental health
Yorkshire and the Humber	• Key high level indicators: GDP and unemployment • "Bundles" of indicators related to strategic objectives • Regional competitiveness indicators (13) • Work with partners especially Regional Assembly's 'Indicator Group'	• Headline 10-year targets (Tier 1) and three-year regional outcome targets (Tier 2) agreed between region and government • 'Tier 3' targets, specifically for Y&H RDA, set by government • Key high level indicators: GDP and unemployment • RES work within context of wider range of targets set in RES, RPG and RSDF; creation of Regional Intelligence Network (Yorkshire Futures) to report annually on progress against this context
East Midlands	• Headline indicators: GDP and unemployment • State of region indicators (12) • Regional competitiveness indicators (13) • A number of performance indicators for each strategic objectives • Subsidiary indicators (to be developed with partners) • Emerging sustainability indicators (to be developed with partners)	• 'Top 20' Index, relating to ambition to enter European 'Top 20' regions cover income, employment, equality and environment • More detailed regional targets, to measure economic performance, cover 12 strands • Wealth and productivity (4); Enterprise (3) Enterprising communities (3); Employment and learning skills (6); Innovation (1); International trade and inward investment (10); Economic growth and the environment (3); Site provision and development (1); Transport (2); ICT (1); Tourism (1);

Table 9.3 (continued)

East Midlands	• Emerging sustainability indicators (to be developed with partners)	Rural development (1); Urban regeneration (2) • Progress will be published annually in State of the Region Report
West Midlands	• Define targets on a number of levels • Headline condition measures linked to regional vision • Thematic condition measures linked to each aim • Output response measures linked to effects of specifications • Some will be same as the State of region indicators, others are region-specific • Monitoring on an annual basis	• Targets to be defined on an number of levels • 'Headline condition measures' linked to regional vision • 'Thematic condition measures' linked to each aim • 'Output response measures' linked to effects of specific actions • Refers to 15 DETR suggested 'condition measures'
North East	• State of region indicators (12) • 16 other indicators • More comprehensive set of targets and indicators may be developed in the future • Work with partners to develop 'Regional Sustainable Development Framework' which will define key objectives, targets and indicators	• Based on government's RDA performance measures: 'Tier 1' objectives; 'Tier 2' regional outcome targets for eleven policy areas; 'Tier 3' milestones or output targets • Specific targets for the North East to be set through RDA's corporate planning process

Source: Baker and Wong, 2001, 2004

According to the above statement, the Regional Competitiveness indicators are contributing factors (i.e. input measures) to the economic development process, while the State of the Region indicators are descriptive indicators to illustrate change (i.e. more related to outcome). While one can agree that the State of the Region indicators tend to focus on the outcome and performance of regional development, it is not that easy to make a clear conceptual distinction between the two sets of indicators. As discussed in Wong's (2000) research, the original Regional Competitiveness indicators were very poorly conceptualised. Instead of measuring the performance of regions, as stated in the 1997 White Paper, DTI has always maintained that these indicators are statistical information that 'illustrates the factors determining regional competitiveness' (i.e. *input* measures) (DTI 1998: section I). Nevertheless, it contradicts itself in the same

document by pointing out that, 'some factors may not determine regional competitiveness but measure outcomes reflecting the competitiveness of a region.' (DTI
1998: Section III). Since then, adjustments and revisions were introduced to the
Regional Competitiveness indicators. In spite of the widely published regional
development and place competitiveness literature (e.g. Amin and Thrift 1994;
Porter 1990), there has been no serious attempt to clarify the meaning of
competitiveness or to establish the causal relationships between certain socio-
economic factors and the performance of a region. It is therefore not surprising to
find that indicators, such as GDP and employment levels, do not sit comfortably
as determining factors when they are, in fact, obvious measures of the outcome
of competitiveness. Without the guidance of an established theoretical framework, a number of key dimensions are used to provide a common-sense model,
and a few indicators are then selected for each of these headings (see Table
9.2). Thus, other than providing the definition and technical details of each indicator, comments are made on neither the comprehensiveness of the framework
nor the adequacy of these indicators.

The above discussion also demonstrates that the decentralisation of regional
decision-making has been followed by a concomitant centralised approach
towards monitoring and evaluation through the issue of guidance notes and the
specification of core indicator sets. It is, however, interesting to note that there
has been a certain degree of political sensitivity of requesting RDAs to compile
performance measures. Hence, right at the beginning of the process, the
emphasis was to encourage RDAs to monitor their progress with a promise that
the indicators would not be used to judge their performance in the first instance.
While significant effort has been made across different government departments
to review, rationalise and harmonise different regional indicator sets to reduce the
problem of information overload, such attempts may sometimes result in the
'lowest common denominator' effect. Following the amalgamation of the two sets of
indicator, the nature of these indicators in relation to regional development is even
more ambiguous than before. Are they input measures or outcome measures?
No one is any wiser from the release of a woolly statement that they are doing
both. Thus rather than underpinning the indicators with a clear conceptual framework, a pragmatic choice of an objective-led framework has been adopted.

FROM DIAGNOSTIC TO PERFORMANCE MEASURES

Reading the lines between the guidance given by the DETR (1999a) and DTI
(1999), RDAs are encouraged to use the suggested core indicators to inform
their policy formulation and to monitor the progress made. In spite of the earlier
promise that the indicators would not be used to judge the performance of RDAs in

1999, the need of performance monitoring began to emerge in 2002. The government introduced economic regeneration performance targets on 9 March 2001 as part of the framework of targets and review in the light of the 2002 Spending Review (DTI 2002b). In April 2002, the Treasury introduced an output and outcome framework to monitor the performance of RDAs. They are regional outcome targets to be delivered with other partners in each region. These targets are mostly derived from national Departmental Public Service Agreement targets. The DTI will agree with each RDA the figures for each regional target as part of the corporate planning process. The RDA will be held accountable for these targets, and will be expected to work with regional partners in developing and delivering these targets. In return, greater financial flexibilities have been given to RDAs in setting the development agenda for the region within the Single Financial Framework (which replaced the previous funding programmes of receiving allocation from each contributing government department).

The performance measures are set within a three-tier targets framework (DTI 2003). Tier 1 is the key national strategic objectives for achieving sustainable economic growth in the long term. These national objectives are achieved through the Tier 2 framework, which includes eleven regional outcome targets (sustainable economic performance, regeneration, urban, rural, physical development, employment, skills, productivity, enterprise, investment and innovation) to be delivered by the RDA and other regional stakeholders over a 3-year period. There are five output milestones in the Tier 3 programme targets that are directly related to the activities and resources of the RDA. In addition, the RDA is expected to set supplementary milestones to reflect the economic circumstances in its region. Hence, these three tier targets are central in the latest monitoring frameworks of Regional Economic Strategies (Baker and Wong 2004). Besides these tier targets, the other additions in the monitoring strategy are the quality of life and sustainability indicators, which reflects the fact that sustainable development has worked its way into the agenda of the RDAs. This means that there is a less distinctive divide between the indicators monitored by the Regional Economic Strategy and the Regional Planning Guidance now than was the case a few years ago.

The 'carrot and stick' approach of decentralising policy-making and centralising performance scrutiny has been a common practice since the last Conservative government. However, when devising some of these performance targets, there is a knowledge gap of whether it is realistic to ask local authorities or regional agencies to collect such data. There is an assumption that the burden of data collection rested on the RDAs and the Regional Observatories, even though certain datasets are notorious for being unreliable at the regional and local levels (see Wong 2000). However, when indicators are no longer simply used for diagnostic purposes but also for performance judgement, the political pressure bites in. Hence, there is a widespread discontent over the

measurement of some of the Tier 2 targets as there is no reliable regional information for certain indicators (e.g. gross value-added data).

The government recently commissioned a major review (Allsopp 2004) of the
information requirements for monetary and wider economic policy-making, with a
more specific remit to assess the demand for, and provision of, regional information. In the report, Allsopp commented that his team was overwhelmed with
submissions from the RDAs over the lack of robust and reliable regional level
data. As discussed in Chapter 4, his report is very critical and makes strong
recommendations on the urgent need to improve such data provision especially
with regard to regional accounts. More importantly, he blatantly argues for more
joined-up thinking in devising performance targets:

> targets need to be measurable, there are benefits from the Office for National
> Statistics or Government Statistical Service experts being involved at an early stage
> of the development of targets, to advise on any associated measurement difficulties.
> All suggestions for the new Public Service Agreement targets, or other government
> targets, should therefore set out how performance can be measured, based on early
> consultation with relevant analysts.
>
> (Allsopp 2004: 18)

On the ground of the critical remarks in the Allsopp Review, there has been
discussion about scrapping the Tier 2 targets.

The review also makes an important comment on the growing demand for
data to measure accurately the output of the government sector to assess the
performance of government departments against their objectives, including
action to raise the productivity of public services as part of the government's
productivity agenda. However, as the report rightly points out, measuring public
services has proven to be difficult and is further complicated by a need to differ-
entiate immediate policy outputs from the wider policy outcomes. It is,
nevertheless, interesting to note that another review led by Sir Tony Atkinson is
already underway to examine ways of improving measure of government output
and productivity.

POLICY DEVOLUTION AND INSTITUTIONAL SUPPORT

As discussed in Chapter 4, the idea of developing regional intelligence capacity
was mooted right at the inception of the RDAs. The proliferation of different
models of regional observatory, though with great variability in resources and
partnership arrangements, and the establishment of the Association of Regional
Observatories, have become the regional infrastructure of policy intelligence.
Nevertheless, seven years on since the publication of the *Building*

Partnerships for Prosperity White Paper (DETR 1997c), the political pressure mounting for better regional statistics has in turn posed a challenge to the existing statistical infrastructure. The Allsopp Review on economic policy and regional information has opened up the Pandora's box of our deficient regional data infrastructure. In the foreword to the report, Allsopp states that 'Devolution of policy responsibility requires changes in the statistical system to go with it – there should be no economic policy responsibility without statistical provision' (Allsopp 2004: 1). The review recommends changes on regional data provision as well as institutional structure. The pressing need for better regional data is pointed out as one of the two major changes required in the statistical system in the UK. The other one is the provision of proper coverage and detail of the service sectors. There are four key areas of recommendation with regard to regional data:

- Bring Regional Accounts more into the National Account framework, including development of an improved and timely measure of real regional Gross Value Added;
- Expanding the range of microeconomic and sub-regional data already available, with the infrastructure used by the Office for National Statistics (ONS) Neighbourhood Statistics Service becoming the primary platform for area-based National Statistics.
- Have an ONS or Government Statistical Service (GSS) presence in the English regions to complement that which already exists in Scotland, Wales and Northern Ireland; and
- Provide greater access for the ONS to administrative data held within government, which could improve both regional and national data while offering important savings in the compliance burden on business.

(Allsopp 2004: 8)

On the institutional front, the review strongly advocates the role of the regional observatories and the Association of Regional Observatories in developing a common data quality platform across all regions to ensure consistency in data collection for comparative analysis. However, it further argues that there is a need to fully integrate its recommendations with the ongoing modernisation programme in the ONS, and that good links should be established between the ONS or GSS statisticians at the centre and those located in the regions and devolved administration. While recognising the importance of regional autonomy, it emphasises that there are significant advantages of having data compiled on a consistent basis. Hence, it recommends the move towards using a common sampling frame for business surveys and suggested the potential role for a 'kitemark' that would indicate a common approach has been adopted according to ONS and GSS guidance.

After the Social Exclusion Unit's *Better Information* report (SEU 2000), the review conducted by Allsopp marks another milestone in taking stock of the UK's data infrastructure for policy-making. The report is remarkable in several counts: it was written in a very critical tone, and Allsopp and his team did not shy away from making some sensitive but far-sighted recommendations to shake up the institutional arrangements of national and regional statistics. In spite of all the justifiable scepticism, the evidence-based policy regime in the UK has succeeded to forge a closer dialogue between data providers and policy-makers, between the centre and the regions, and between different government departments. It is the political-managerial requirements of better information that raise the profile of the technical capacity of data infrastructure at different spatial levels.

CONCLUDING REMARKS

This chapter has provided an overview of the changing policy context of indicators usage in the urban and regional development field, recent progress in academic and consultancy research on quantitative measures, and the application of indicators in urban and regional development policies. Indicators measuring quality of life and place competitiveness have gained a prominent position in the worldwide sustainability movement and the current evidence-based policy regime in Britain. Although a more positive atmosphere, at least in Britain, has been evident over indicators development and their application in policy-making, it is clear that many of the problems encountered in developing indicators have not been resolved. If anything, the mass consumption of policy indicators simply exposes the inadequacy of the data infrastructure to support such an information-intensive policy regime and the lack of joined-up thinking in devising monitoring guidance and performance targets. On the positive front, noises and criticisms of such technical matters have at last found their way into the public arena to lobby for attention and resources. This is as yet too early to make a judgement on the changing culture of using quantitative information to formulate and monitor urban and regional policies. It is, therefore, important to keep an eye on how indicators affect both the decision-making process, and the outcomes of decisions made, in the longer term.

SUSTAINABILITY AND PLANNING INDICATORS

WICKED PROBLEMS AND OPEN CONCEPTS

Policy problems such as environmental and economic issues that planners have to deal with can be seen as 'wicked problems'. According to Rittel and Webber (1973), there is a set of policy problems that cannot be resolved with traditional linear analytical approaches. This is because these problems tend to be found in an evolving set of interlocking issues and constraints. Each attempt to create a solution may reveal another, more complex problem. This means that formulating the problem and the solution is essentially the same thing, and there is no definitive statement of the problems. Another important characteristic of wicked problems is that they are always embedded in a dynamic social context, which makes each problem unique. Within this broad context, there are many stakeholders who are, nevertheless, interested in resolving the problems. However, the social complexity of these problems makes it difficult to achieve consensus over whatever solutions emerge.

To Rittel and Webber, the classical systems approach based on the assumption of understanding the problems, gathering information, synthesising information and working out solutions will not work for wicked problems, because it is not possible to search for meaningful information without grasping the nature of the problem. Hence, the appropriate way to tackle wicked problems is to develop shared understanding and shared commitment. Consensus emerges through the process of laying out alternative understandings of the problem, competing interests, priorities and constraints. Only when the problem is framed and articulated in a concise and well-bounded manner, is it possible to apply more formal analytical tools. In this sense, wicked problems have the same traits as what Greer (1969) called 'the changing nature of problem definitions'.

Following Greer and Rittel and Webber's arguments, it is clear that developing indicators to measure planning concepts is going to be a major challenge. According to Innes (1990: 126), the iterative process of formulating these open concepts can require at least ten or fifteen years of trial and error to make them workable. In the light of the global movement of sustainability, and the newly introduced planning reforms in England, this chapter explores the process of developing sustainability and planning indicators where the technical complexity of measurement meets with the social complexity of politics and competing interests.

GLOCALISATION OF SUSTAINABILITY INDICATORS

Following endorsements of national governments in the 1992 United Nations Conference in Rio de Janeiro, the international action plan 'Agenda 21' urged that 'indicators of sustainable development need to be developed to provide solid bases for decision-making at all levels' (UNCED 1992: Chapter 40). In spite of the general acceptance of the widely quoted definition of the World Commission on Environment and Development that sustainable development is 'development that meets the needs of the present generation without compromising the ability of the future generations to meet their own needs' (World Commission on Environment and Development 1987: 8), different commentators and organisations have used the concept in different ways. Its interpretation is dependent on philosophical considerations that are influenced by political, ethical, religious and cultural factors (Schaller 1993). Bell and Morse thus queried 'how can something so vague be so popular' (Bell and Morse 1999: 9)? Their findings show that the uncertainty over the meaning of sustainability has not reduced its popularity. Indeed, it is exactly the vagueness and flexibility of such an open concept, as Greer suggested, that makes it so attractive to different stakeholders and thus remains in the mainstream. Although there is a general consensus that sustainability should encompass social equity, economic growth and environmental protection, the emphasis attached to these different aspects of development varies from definition to definition. The development of indicators measuring sustainability adds an extra dimension of ambiguity, as sustainability indicators can be classified by their functions and roles in the decision-making process. Some aim to provide a simple description of the current state of development (state indicators), while others are used to diagnose and gauge the process that will influence the state of progress towards sustainability (pressure, process or control indicators), or to assess the impact brought by policy changes (target or performance indicators).

One interesting aspect of the sustainable development agenda is that while it is very much an international institutional-led initiative of the United Nations, it also has tremendous appeal at the local community level. The pledge is, in fact, directed at the local level in Chapter 28 of the document. There is a recognition that the delivery of action, based on the Agenda 21 policies, relies heavily on local partnerships that involve local government, business and voluntary sectors,

though it is clear that local authorities are seen as providing the right level of governance to 'construct, operate, and maintain economic, social and infrastructure,
oversee planning processes, establish local environmental policies and regulations,
and assist in implementing national and subnational environmental policies'
(UNCED 1992: Chapter 28.1). Agenda 21 is thus seen as providing a common
voice for local government worldwide to strengthen local sustainable development
planning, which in turn assists in the global implementation of Agenda 21 policies.

Under the canvas of this exciting worldwide ideology, sustainability indicator
sets are rapidly developing as the progress of achieving the goals embraced in
Agenda 21 have to be measured and calibrated to produce environmental
audits and assessment. The importance of sustainability indicators is clearly
articulated in Chapter 40 of Agenda 21, which specifically calls for harmonisation of efforts to develop indicators at the national, regional and global levels.
This global–local interplay over Agenda 21 bears implications for the development of sustainable indicators. From the global perspective, the United Nations
Commission on Sustainable Development (UNCSD 1996) has taken a lead by
publishing the *Indicators of Sustainable Development: Framework and
Methodologies* in August 1996. The publication contained methodology
sheets for 134 indicators of sustainable development under four primary dimensions: social, environmental, economic and institutional. The 'pressure, state,
response' link model (Hammond *et al.* 1996) was used to conceptualise the
chain effect of human activities on the changing state of our environment and
resources. Based on voluntary testing in twenty-two countries and expert-group
consultation, a revised set of fifty-eight indicators (under fifteen themes and
thirty-eight sub-themes) and methodology sheets were included in the
Indicators of Sustainable Development: Guidelines and Methodologies
(September 2001). The report was prepared as the culmination of the 5-year
Work Programme. The methodology sheets provided a description of each of
the indicators, its policy relevance, underlying methodology, data availability
assessment and sources. The testing of the indicators has received support from
Eurostat, which led to the publication of the UNCED indicators for the
European level (Eurostat 2001b).

The process of testing the indicators was found to be a positive catalyst for
cross-national collaboration in developing the indicators as well as advancing
the goals of sustainable development (UNCSD 2001). However, some
constraints, especially institutional mechanisms, were identified as problematic
during the process:

> limitations on the availability of finance and human resources; difficulty in mobilizing
> the relevant experts and stakeholders; lack of coordination between statistical agen-
> cies and the indicator focal point; low level of awareness among stakeholders, low

level of commitment on the part of participating institutions; competing work demands and government leadership transitions that resulted in discontinuities in the imple- mentation in the indicators process.

(UNCSD 2001: 9)

Following the publication of the Indicators of Sustainable Development, the UNCSD has been planning to develop diagnostic vulnerability indicators to help monitor the implementation of National Sustainable Development Strategies. The proposed methodology is divided into three areas: national performance, capacities and opportunities. The effort put in developing sustain- able indicators by the United Nations shows that this global strategy is strongly based on the rational approach of policy-making. As clearly indicated by one of its officials,

> The guidance document underscores the need to anchor the strategy process in sound technical analysis. Putting in place an effective monitoring and evaluation mechanism is vital for the strategy process. . . . Monitoring and evaluation needs to be based on clear indicators and built into strategies to steer processes, track progress, distil and capture lessons, and signal when a change of direction is necessary. These indicators could be both qualitative and quantitative, and should reflect the status and trends of a particular process element or product.
>
> (Shah 2004: 4)

Another interesting observation is that, although local communities are empow- ered to develop their action plan to achieve Agenda 21, the approach towards the development of sustainable indicators is by and large a 'top-down' one. As the United Nations openly stated in its 2001 report,

> the primary goal of the indicator programme, however, is to develop a means to assist national decision-making. On the other hand, it is considered that a good indi- cator system should be able to reflect the specific issues and conditions of a country or a region but should nevertheless be harmonized internationally to the extent possible.
>
> (UNCSD 2001: 10)

This raises the thorny issue of harmonisation among different indicator sets.

In Europe, parallel efforts have been made to develop sustainability indica- tors as well as harmonising local indicator sets. Through the Communication on 'Sustainable urban development in the European Union: a framework for action' (European Commission 1998), the European Commission urged the importance of integrating local sustainability into its policies and to monitor the progress made on Local Agenda 21. At the supra-national level, European institutions such as the European Environment Agency and Eurostat have been defining and collecting indicators to develop Environmental Indicators, Environmental Pressure Indices and the Urban Audit. However, these aggregated indices, primarily used at the national level, were not found to be so helpful at regional and local levels in informing the progress of sustainable development. Hence sustainability indicator sets, comprising a broad range of indicators, are rapidly emerging in local communities across Europe (e.g. Audit Commission and the Local Government Management Board in the UK, Ecosistema Urbano in Italy) (Ambiente Italia 2003). In May 1999, The European Commission took a lead towards harmonising different local indicator sets through the European Common Indicators initiative. This initiative involves a partnership of different organisations and levels to find comparable data and gain a better understanding of sustainability in local communities across Europe. The rationale that underpinned the European Common Indicators is to integrate local action towards sustainability by providing complementarily to, rather than replacing, existing local, national and sectoral indicator sets by following the principle of subsidiarity (Ambiente Italia 2003). Ten common local sustainability indicator groups have been identified out of a list of 1,000 potential indicators through a bottom-up process and the first set of data became available in autumn 2001 (European Commission 2000c). Between January 2001 and February 2003, a 2-year testing project was carried out with twenty-five participating local authorities. On the basis of suggestions and proposals from the participants, a document with all methodology refinements has been drafted and a headline indicator has been chosen for each of the ten European Common Indicator Groups (see Table 10.1).

In the United Kingdom, the government has signed up to the sustainability agenda, through the 1990 White Paper *This Common Inheritance* (HM Government, 1990) and the subsequent *UK Strategy for Sustainable Development* (DoE 1994), which was, itself, a response to Agenda 21. This new environmental agenda has brought with it a need to employ indicators as a key mechanism for assessing environmental impact and capacity (Maclaren 1996; Macnaghten *et al.* 1995). Subsequent enthusiastic responses to Agenda 21 can be found in nearly every published report (e.g. Countryside Commission *et al.* 1993; DoE 1993; Arup Economics and Planning 1995). This blossoming of sustainability indicators, however, had not been subject to as much government co-ordination across the national, regional and local spectrum as might have been expected (Stewart, 1995a, 13). The critical step towards greater harmonisation was, in fact, first taken by the Local Government Management Board (LGMB 1995), which carried out a full study on sustainability indicators at the local level, including pilot projects in six local authority areas. However, the conceptual framework within the LGMB report did not offer any new thinking on the definition of

Table 10.1 The European Common Indicators (Issue Groups and Headline Indicators)

• Citizen satisfaction with the local community   Headline indicator: Average satisfaction with the local community
• Local contribution to global climate change   Headline indicator: CO_2 emission per capita
• Local mobility and passenger transportation   Headline indicator: Percentage of trips by motorized private transport
• Availability of local public open areas and services   Headline indicator: Percentage of citizens living within 300m from public open areas > 5,000 m2
• Quality of local air   Headline indicator: Number of PM10 net overcomings
• Children's journeys to and from school   Headline indicator: Percentage of children going to school by car
• Sustainable management of the local authority and local businesses   Headline indicator: Percentage of environmental certifications on total enterprises
• Noise pollution   Headline indicator: Percentage of population exposed to L night > 55 dB(A)
• Sustainable land use   Headline indicator: Percentage of protected area
• Products promoting sustainability   Headline indicator: Percentage of people buying "sustainable products"

Source: Ambiente Italia, 2003: 167.

sustainable development, and the thirteen identified themes only covered very general aspects of sustainability (Stewart 1995a). The result of the pilot projects showed that there were great variations over the interpretation of sustainability at the local level, and it was very difficult for local communities to relate social aspects such as local needs, and more abstract issues like aesthetics, to the concept of sustainability. The involvement of local community groups was found to be valuable, but there were difficulties in getting certain groups to participate, especially the business sector. Finally, it was interesting to note that altogether ninety-five indicators were chosen by the six pilot projects, only seven indicators were commonly selected by more than three of them.

The government's attempt to harmonise sustainability indicators did not come until 1998 through the publication of the *Sustainability Counts* consultation document (DETR 1998a). As discussed in Chapter 9, a set of thirteen headline indicators was selected out of the national set of 120 indicators. This brave attempt at information reduction was partly a response to its earlier recognition in the *Indicators of Sustainable Development for the United Kingdom* report (DoE 1996) that a more limited number of core indicators would be needed in the future to avoid information overload. Information on the headline indicators, now known as Quality of Life Count headline indicators, is provided for the nine English Regions and for Wales. The national indicator set currently contains 147 indicators and a subset of 15 key headline indicators (DEFRA 2004). As discussed earlier in Chapter 9, there is currently discussion between the government and the Audit Commission on the need for harmonising the two different sets of quality of life indicators.

The 1996 Sustainability Indicators report also touched upon the issue of recalculating the national indicators at the local level to allow local comparison with the national 'norm'. The guidance was, however, somewhat ambiguous at that time: although the government agreed that national indicators might not adequately reflect local circumstances, and encouraged local communities to develop local indicators, it issued the warning that,

> if every local area develops indicators in its own way for its own use, all defined and constructed in different ways, no overall national or regional picture can be obtained, nor is it possible for an area to compare its local situation or progress.
>
> (DoE 1996: para. 2.19)

In order to provide a framework to guide indicator development at the local level, in May 1998, the government and the Improvement and Development Agency in consultation with the Local Government Association set up the Central–Local [Government] Information Partnership (CLIP) Task Force on Sustainable Development. A core menu of twenty-nine indicators (including Quality of Life Count headline indicators) was recommended for local authorities through the launch of the *Local Quality of Life Counts Handbook* in July 2000 (DETR 2000a). This was the result of an extensive consultation with over 100 representatives from local authorities, Local Agenda 21 groups and others.

The discussion here highlights the fact that there is not a single perfect measure of the complex concept of sustainability, and sustainability indicator sets have been spawned at different spatial levels. The blossoming of local sustainability indicator sets has provided flexibility for local communities to identify issues that reflect their particular concerns and circumstances. The problem is that these indicator sets are not necessarily compatible, and the diverse nature of these indicators makes it impossible to undertake meaningful benchmarking and comparison of progress across different spatial scales. There is also an articulated fear that the failure of local communities to grasp the abstract concept of indicators, and certain aspects of sustainability, will lead to the sidelining of these components in the sustainable development agenda. This means that there is an inherent tension between local specificity and global universality in the process of developing sustainability indicators.

The bewilderment of information overload clearly demonstrates the need to harmonise different indicator sets. It is, nevertheless, interesting to note that the harmonisation process is taking place concomitantly across different levels, but largely led by the higher-level governing institutions. In spite of their emphasis on using a bottom-up approach of consultation and representation, it is still largely a top-down process. While the United Nations' concern is to harmonise indicators at the national level, both the EU and the UK government are interested in sorting out the local level. When there are concomitant efforts in harmonising indicators, both vertically across different spatial scales and horizontally across different policy sectors and different organisations, is harmonisation still meaningful? The irony is that, after all these concerned efforts, we may still have a whole array of core/ headline/common/flagship indicators produced by different institutions competing for their own cause. The development of sustainability indicators has manifested the interplay of global–local politics in shaping and casting the development of technical indicators. It has also proved that the wicked spell is already taking effect; the solution to one set of problems has triggered another set of problems.

FROM SUSTAINABLE DEVELOPMENT TO SUSTAINABLE COMMUNITIES

This section aims to examine how the philosophy and nature of sustainability indicators has developed and evolved. The discussion will first focus on the experience in the USA and then on recent developments in British policy circles.

THE FOLK MOVEMENT OF COMMUNITY INDICATORS IN THE USA

Increasing concerns over quality of life issues and the sustainability movement has created a culture of making use of indicators as a vehicle to understanding and addressing community issues in the USA since the mid-1980s. This is widely known as the 'community indicators movement' and various commentators (e.g. Innes and Booher 2000; Sawicki 2002; Swain and Hollar 2003) have written academic research papers that chart the approaches used and the rationale that underpinned these indicators. More importantly, critical suggestions have been made on the strategies required to improve the quality and the usage of these indicators. In spite of the burgeoning of community indicator sets, they have different origins, purposes and assumptions. In spite of the slightly different classifications by Sawicki and Swain and Hollar, the measurement of community indicators tends to revolve around three sets of concern: community well-being (quality of life and liveability studies); sustainable development (sustainability projects with an emphasis on environmental issues); and government performance (policy outcome and performance measures and benchmarking projects).

One interesting observation is the distinction between quality of life and sustainability indicators even though they are closely related. As Liu (1976: 7–8) pointed out, the quality of life concept was regarded as a potential new tool for decision-makers and, interestingly, his research study was sponsored by the US Environmental Protection Agency in 1972. Despite the common ground shared by the two concepts, researchers in the USA are very keen to demarcate the differences between the two concepts and their respective indicator sets. Sustainability indicators are seen to have much stronger requirements from the stakeholder to buy into a particular way of thinking and rely on an ecological frame of reference to reveal the interconnectedness of social and physical worlds over time and across space (Sawicki 2002; Schuessler and Fisher 1985; Swain and Hollar 2003). Quality of life indicator projects, on the other hand, tend to start with the task of goal and value definition, which allows the community to freely articulate their visions and agreed parameters of what constitutes quality of life (Swain and Hollar 2003).

The Jacksonville Community Council Inc. (JCCI) Indicators Project is the pioneer and leader in the community indicators movement (Swain and Hollar 2003). The project started in 1985 with financial support from the Jacksonville Regional Chamber of Commerce and over 100 volunteers. From the introduction of the JCCI web page, it is clear that the focus of the project is about community engagement and building better quality of life: 'JCCI is a nonpartisan civic organization that engages diverse citizens in open dialogue, research, consensus building and leadership development to improve the quality of life and build a better community in Northeast Florida and beyond' (JCCI 2005: web page). While environmental quality is part of the project, the conviction of JCCI is on measuring the wider community progress. Besides extensive citizen participation and business finance, the project also has in-house research capacity to serve as a community 'think tank' (Swain and Hollar 2003).

Another widely quoted successful example is Sustainable Seattle (Bell and Morse 1999; Sawicki 2002). The mission of Sustainable Seattle is to develop

Indicators of Sustainable Community to improve the central Puget Sound region's long-term health and vitality. Due to the rapid growth of the region into a large metropolitan area, the local community has a strong concern over the quality of life in some neighbourhoods as well as the health of the Puget South ecosystem (Sustainable Seattle 2005). A comprehensive set of forty indicators is used to track the sustainability trends of the region. Since its inception in 1993, the project closely followed the sustainability principle of the WCED and the indicators have to look at the inter-connectivity among ecological, social and economic factors, as well as addressing the inter-generational nature of those factors. The quality of life in the region is thus closely wrapped up in the framework of ecological sustainability.

While the approaches to community indicators have their specific emphasis and perspective, they are not mutually exclusive in practice. For instance, state and municipal governments have been involved in tracking progress towards quality of life and sustainable development, and used such indicators to monitor performance of public services and guide budget decisions. The most fascinating aspect of this phenomenon of community indicators development is, however, the extensive citizen involvement and financial support from the business sector to establish long-term community-based projects.

CENTRALLY DIRECTED COMMUNITY STRATEGIES IN BRITAIN

The trajectory of the development of sustainability indicators in Britain is somewhat different from the situation in the USA. First of all, central government has played a central role in shaping and guiding the development of sustainable indicator sets. In the process of doing so, it also defines and redefines the meaning and interpretation of sustainable development. When preparing for the *Indicators of Sustainable Development for the United Kingdom* report (DoE 1996), the Working Group abandoned the idea of adopting the elaborate 'state–pressure–response' model because of its complexity and opted to separate out indicators concerning the economy, the environment and the actors involved (Cannell *et al.* 1999). In spite of its subsequent effort in rationalising the indicator sets, the conceptual framework underpinning the selection of headline indicators remains problematic. Although the indicators were selected on the basis of the four sustainability objectives encompassing social, economic and environmental aspects of development, it is not clear how they inter-connect with each other in order to contribute to the central concern of sustainability (DETR 2000a; Levett 1998). As explained in Chapter 9, the change of branding from *sustainability* to *quality of life* count indicators provides the government with some flexibility over its interpretation of the meaning of these indicators. One can see that there is a shift in the emphasis from ecological environmental concerns to the broader meaning of quality of life, though still based on the principles of sustainable development. This is explicitly stated in the *Local Quality of Life Counts Handbook*:

> Quality of life is a term used by government synonymously with sustainable development, because it is felt to be more easily understood by the general public. However, care needs to be taken in using it – quality of life for people today must not be achieved at the expense of people in the future.
>
> (DETR 2000a: 3)

Another major development is to replace the non-statutory, spontaneous movement of Local Agenda 21 with the statutory *community strategies* (DETR 2000a). Local authorities are given the new duty of preparing community strategies to improve the social, economic and environmental well-being of their areas to achieve sustainable development in the UK. The guidance from the government is that 'the duty to produce a Community Strategy is similar to the process of producing strategies under Local Agenda 21' (DETR 2000a: 3). The issue of the fifteen local quality of life count headline indicators aims to facilitate the process of preparing community strategies, and the guidance in the handbook intends to offer 'some ideas for measuring sustainable development and quality of life in local communities' (DETR 2000a: 3). Hence, the process to harmonise local indicators is actually part of the process to streamline and consolidate the policy framework of a wide range of local initiatives such as Local Agenda 21, local transport plans, local crime and disorder strategies, health improvement plans and local environment strategies or plans. The irony is that the spontaneous action from Local Agenda 21 partnerships has been replaced by centrally controlled, government-led, 'community' strategies where citizen and business participation is not as extensive and enthusiastic as that witnessed in some projects in the USA.

The discussion here shows that the term 'quality of life' functions in what Schuessler and Fisher (1985: 131–2) called 'a meta-theoretical way' to suit the government's policy agenda in Britain. While in the USA, researchers can broadly identify different approaches to community indicator development, the situation in Britain tends to be rather uniform. Community indicators are just another set of administrative tools to be included in the package of developing community strategies. Hence, such indicators could be seen as a product, created by local government and its partners, for the community under a framework established by Whitehall.

MORE INDICATORS FOR SUSTAINABLE COMMUNITIES: THE EGAN

REVIEW

The jargon of sustainable development took a further twist in 2003 after the British Government launched the *Sustainable Communities: Building for the Future* document (ODPM 2003a) which sets out a long-term programme of action and investment in housing and planning with the aim of building thriving sustainable communities. This includes substantial investment in housing improvements and the provision of affordable housing, as well as a particular focus on four identified major growth areas located within London, the South East, the East and (the southern part of) the East Midlands. Proposals for a Northern Growth Corridor, stretching from Liverpool to Hull and Sheffield to Newcastle, were subsequently added in the *Creating Sustainable Communities: Making it Happen: the Northern Way* (ODPM 2004a) report. The term 'sustainable communities' has now become the vogue policy vocabulary, so much so that the government's main policy statement on the operation of the planning system (PPS1) is termed 'Creating Sustainable Communities' (ODPM 2004b) in its consultation draft, though it was reverted back to 'Delivering Sustainable Development' (ODPM 2005c) in its final version. It is clear that 'sustainable communities' is now being used interchangeably with 'sustainable development'.

While 'sustainable communities' has become the buzzword in planning and housing circles, there are concerns that the concept is too broad and loosely defined. Hence, a research report was jointly commissioned by the Economic and Social Research Council and the Office of the Deputy Prime Minister to elicit the exact nature and meaning of sustainable communities. In the report, Kearns and Turok defined sustainable communities as:

> settlements which meet diverse needs of all existing and future residents; contribute to a high quality of life; and offer appropriate ladders of opportunity for household advancement, either locally or through external connections. They also limit the adverse external effects on the environment, society and economy.
>
> (Kearns and Turok 2003: 3)

They, however, emphasised the need to further deconstruct the concept to enhance 'deeper understanding, assembly of evidence and practical policy formulation' (Kearns and Turok 2003: 3). Ten indicative criteria and sixteen principles and values were identified to assess the sustainability of communities, which resulted in six broad components of sustainable communities, namely: residential environment, economy, local services, transport and connectivity, society and culture, and governance. A presentation based on the report was made in June 2003 to the Egan Task Force. The Task Force, headed by Sir John Egan, was asked by the Deputy

Prime Minister to conduct a review to consider the skills needed to help deliver the vision and aims of the Sustainable Communities Plan.

The Egan Review stressed the importance of creating a 'common language' to help provide a benchmark against which progress towards sustainability can be measured (Egan 2004: para. 1.7). He defined sustainable communities as ones that 'meet the diverse needs of existing and future residents, their children and other users, contribute to a high quality of life and provide opportunity and choice' (Egan 2004: para. 1.3), which was largely based on Kearns and Turok's definition. The seven key components of sustainable communities identified in the report also resemble Kearns and Turok's list, though the residential environment component is split into environmental, and housing and the built environment components. The creation of common language is not just about defining sustainable communities, but also relates to the specification of associated monitoring indicators. Over 400 existing indicators, including those from the Audit Commission, the Office of the Deputy Prime Minister and the Department of the Environment, Food and Rural Affairs, were considered by Egan and his team. A total set of fifty indicators, between four and nine indicators for each key component, has been recommended to local authorities for monitoring progress towards sustainability. What is interesting is that, rather than simply select the indicators from existing ones, the report recommends four new/piloted indicators. This is partly due to the fact that the review team considered the use of subjective indicators, to reflect the perception of residents, as part of the process towards delivering sustainable communities. Hence a mixture of subjective and objective data is included in the recommended indicators set.

It is also interesting to note that the Egan Report strongly emphasised that local authorities and their partners should exercise their judgement in selecting the most relevant indicators to suit their local circumstances rather than treating the proposed indicators as a 'tick box' exercise. Meanwhile, it also stressed the importance of using the indicators as the basis for comparisons, especially international comparisons. The expectation is that, by the end of 2005, local authorities should:

- incorporate in their Sustainable Community Strategies a process through which they and their partners will select the indicators [from the recommended set] that are most relevant to the needs of their communities;
- identify mechanisms for establishing baselines and regularly tracking progress towards achieving sustainability with the longer term aim of tracking all of the indicators; and
- make provision for taking action to address poor performance where it occurs. Feedback to local people should be an essential part of this process.

(Egan 2004: para. 1.16)

The report also urged central government departments to recognise the importance of these indicators, and their use at the local level, alongside their existing performance targets.

The discussion above clearly demonstrates the popular appeal of the ideas and principles of sustainable development. These principles have been readily embraced into the process of devising strategies for community development. However, there is not a consensual approach as to how far the ecological roots of sustainability are embedded into the strategies. Hence, some of these strategies opt to lean on the more simple and vague approach of promoting quality of life. As Hoernig and Seasons observed, 'in contrast to both sustainability and healthy cities approaches, [q]uality-of-life indicators tend to take a 'bricolage' approach that clumps together a number of available indicators across a range of sectors' (Hoernig and Seasons 2004: 88). This is definitely the case in Britain. The policy discourse of sustainable development has been shifting, and the terminologies of sustainable development, quality of life and sustainable communities are being used synonymously. Meanwhile, continuous effort has been made to clarify and refine the definition of these terms in guidance documents, usually accompanied by a parallel set of proposed monitoring indicators. In spite of the shared political desire to harmonise indicators and use existing data sources, there are always some additional new ones proposed. In a way, this is inevitable, as different analysts will have a different viewpoint of what indicators best suit a particular interpretation of their definition of sustainable development.

MONITORING SPATIAL PLANNING STRATEGIES

The proliferation of indicator sets for monitoring local and regional planning practice is noticeable over the last decade. Hoernig and Seasons (2004) provided a detailed account of such practice, especially in North America. On this side of the Atlantic, the wholesale reform of the British planning system in autumn 2004 marks the beginning of a new era in developing 'spatial' rather than purely land-use plans. The new system of Regional Spatial Strategies and Local Development Documents adopts a spatial planning approach that will 'integrate policies for the development and use of land with other policies and programmes which influence the nature of places and how they function' (ODPM 2004b: 13). Another significant shift from previous practice is the emphasis on a systematic approach towards strategy monitoring. *Planning Policy Statement 11* explains that the purpose of a Regional Spatial Strategies should be to 'provide a clear link between policy objectives and priorities, targets and indicators. It should be monitored annually against the delivery of its priorities and the realisation of its vision for the region' (ODPM 2004c: para 1.7). A stronger tone is used in *Planning Policy Statement 12,* which states that 'local development documents must be soundly based in terms of their content and the process by which they are produced. They must also be based upon a robust, credible evidence base' (ODPM 2004d: para. 1.3). Such an evidence-based approach to spatial planning has already also taken place at the European level through the establishment of the European Spatial Planning Observation Network (ESPON) Programme. The discussion in this section thus focuses on the ESPON initiative and the monitoring arrangements brought by the planning reforms in Britain.

ESPON: POLITICAL RATIONALITY OR TECHNICAL JUSTIFICATION?

The evidence-based policy-making ethos is strongly embedded in the development and monitoring of European spatial policy. The ESPON Programme was launched after the preparation of the European Spatial Development Perspective (ESDP) in May 1999 in Potsdam to provide a solid analytical base for the ESDP and to address the gaps in comparative, quantified and geo-referenced data. The programme is implemented under the framework of the Community Initiative INTERREG III. The ESPON 2006 Programme was launched by the European Commission in June 2002 to provide a research and intelligence function to serve the Commission's spatial policies with an enlarged EU territory of twenty-five Member States. The scope of activities of the ESPON Programme (ESPON 2005) includes:

> a diagnosis of the principal territorial trends at EU scale as well as the difficulties and potentialities within the European territory as a whole;
> a cartographic picture of the major territorial disparities and of their respective intensity;
> a number of territorial indicators and typologies assisting a setting of European priorities for a balanced and polycentric enlarged European territory; and
> some integrated tools and appropriate instruments (databases, indicators, methodologies for territorial impact analysis and systematic spatial analyses) to improve the spatial co-ordination of sector policies.

As van Gestel and Faludi (2005) commented, the original observatory, based on a Geddesian approach of survey-before-plan, has evolved into a dynamic research network. They were, however, rather critical towards the ways research and intelligence had been deployed by the Commission. They found that the Commission exerted pressure on ESPON to produce results as and when needed to suit its purposes. There is a tension between the technical-rationality of policy-making and the reality of selective use of data to justify political decisions. They highlighted the fact that the Commission desired to have scientific evidence to inform funding allocation and policy decisions, on the one hand, but was keen to retain control over the derivation of funding allocation criteria, on the other. Hence, in spite of the success in carrying out systematic research and compiling a very rich seam of information, many researchers were disappointed in the process. As a result, van Gestel and Faludi suggested that these concerns should be brought into the open, and called for a better understanding of the mutual roles that ESPON researchers and the Commission should play.

MONITORING REGIONAL PLANNING POLICIES: NATIONAL OUTPUT INDICATORS AND CENTRAL GUIDANCE

In many ways, the monitoring of planning policies has been deeply embedded in the British planning system as there has been periodic review of development plans by local planning authorities since the 1970s. While evaluation and monitoring were carried out in the past, they have been more successful in ascertaining the efficiency of the planning system and less successful in establishing its effectiveness (Houghton 1997; Morrison and Pearce 2000). As Houghton asserted, performance indicators tended to focus on procedures rather than outcomes, and often only included expenditure data and administrative statistics on the number of planning applications and appeals processed. He also took a rather cynical view of the reason for an absence of planning performance evaluation, 'since it might well reveal differences between policy intentions and claims of success, and actual outcomes' (Houghton 1997: 9). His remarks are probably correct; however, the full picture behind this lack of assertive evaluation is also due to the nature of planning policies and the practical issues associated with evaluation. A number of issues can be identified:

- The lack of consistent and relevant information: past research finds it difficult to establish the impact and performance of the planning system because relevant information is not available consistently across the board.
- The complexity of multiple influences: while spatial planning provides a framework to achieve the objectives of sustainable development, its delivery is heavily reliant upon other actors and agencies across different policy sectors. Since the outcome of change is brought by the interaction of different forces and different actors in the open system, it is not easy to isolate or break down the impact brought by the planning system. This means that there is a need to develop an analytical approach to ascertain

the impact brought by planning policies rather than accepting the simplistic notion of gauging global change.

- The measure of indirect impact: planning policy can indirectly influence the attitude and behaviour of different actors in the process of environmental change. It is, however, difficult to quantify such indirect impacts. This is due to the fact that it is impossible to establish the counterfactual situation of what would have happened without the planning system.
- The choice of suitable timescale: different aspects of the planning system may require different lead-in times to take effect. For instance, the procedural aspects of change will probably take place earlier than the actual policy outcomes, as the latter requires a reasonable length of time for the policy to take effect and start making an impact.

In spite of these problems, the requirements for planning bodies to monitor the performance of their policies have been tightened since the inception of the Regional Planning Guidance in the late 1990s. Regional Planning Bodies were asked to produce an annual monitoring report on their policies. At the beginning of the process, all regional monitoring frameworks tended to uniformly adopt the DETR's *Quality of Life (Sustainability) Headline Indicators* (Baker and Wong 2001). However, there have been some recent changes in their monitoring frameworks. By 2003, Regional Planning Bodies tended to adopt a more fine-grained indicator framework, to measure the progress made in different policy sectors, than before. This is partly attributable to the publication of a good-practice guide by the Office of the Deputy Prime Minister (ODPM 2002a). This guidance set out some basic principles over the approaches used to monitoring planning policies.

First of all, a core set of fourteen national output indicators were identified in the guidance document (ODPM 2002a: section 6.3). These indicators cover the key sectoral issues on: economic development, housing, transport, regional services, minerals, waste, coastal and river management, biodiversity, and energy. The guidance also emphasised the importance of including contextual indicators to provide the basis for an assessment of the broad socio-economic trends, against which the outputs can be assessed. Second, an 'objectives–targets–indicators' approach was used as the framework for monitoring regional planning guidance. This approach focuses on policy objectives, based on which relevant indicators for monitoring will be selected. This helps to avoid the 'information/indicator-led approach' of monitoring, which collects and evaluates a wide range of information that is not relevant to the performance or specific contribution of planning policies. The feedback from the monitoring process also helps to pinpoint specific performance issues to inform policy-making. Third, there was a requirement to set performance targets for the indicators. The guidance outlined the

'SMART' (Specific, Measurable, Achievable, Realistic and Time-bound) approach in setting such targets.

The issuing of guidance from the centre signifies the evolving vertical relationships between the development of regional monitoring approaches and central government, especially in the context of the ODPM's national output indicators. The reference to the potential formation of monitoring groups in the ODPM's good-practice guide, however, introduces another side to the issue of vertical linkages – those downwards towards the policy development of sub-regional and local actors, particularly local authorities. Indeed, advice in the ODPM guide suggested that such a regional planning monitoring group 'needs to be supported by and will for the most part be dependent upon the monitoring activity of individual local planning authorities' (ODPM 2002a: 3.4.1). Despite this suggested importance of local monitoring activities, it is noticeable that only a few of the Regional Planning Guidance documents mentioned such arrangements (Baker and Wong 2004).

The requirements of policy monitoring have been further strengthened following the latest planning reforms. Regional Planning Bodies are now required to prepare the Regional Spatial Strategies, which have become part of the statutory development plan system. Under the new arrangements, Regional Planning Bodies will need to provide details of their proposed monitoring arrangements to the Secretary of State (or set them out in any draft revision of Regional Spatial Strategies) who will 'need to be satisfied that the Regional Planning Body has established a monitoring and review mechanism, with member local authorities and other bodies as appropriate, that can respond sufficiently quickly to any adverse impacts of the strategy' (ODPM 2003b: 3.3). The consultation paper on monitoring regional strategies also advised Regional Planning Bodies to draw upon the earlier good-practice guidance on monitoring (ODPM 2002a) and a proposed set of national core output indicators was included in the annex to replace the earlier set. After ongoing consultation with Regional Planning Bodies and other key stakeholders, a finalised set of national core output indicators was published in March 2005 (ODPM 2005a).

PLANNING REFORMS AND EVIDENCE-BASED POLICY-MAKING

Following the enactment of the Planning and Compulsory Purchase Act (HM Government 2004), local planning authorities have to produce the statutory Local Development Documents as part of their Local Development Frameworks (LDF). The Act also requires statutory annual monitoring reports that directly link monitoring to the implementation of policy, underlining the importance of monitoring in terms of successful local spatial policies. Critically, authorities are required to explain in their reports what steps they will take if LDF policies are shown to be underperforming, or if framework preparation is not in line with the timetable set out in the local development scheme. Previously, monitoring has been regarded as an error-correcting mechanism to bring land use plans back on track by addressing negative feedback. However, the latest emphasis on monitoring is seen as part of the planning reforms, especially as the Treasury has allocated a significant sum of Planning Delivery Grant to local authorities to deliver a step change in the planning system. Given the ethos of evidence-based governance, there is a need to prove that spatial policies are based on robust evidence and that the policies are effective in delivering the government's objectives of creating sustainable communities and economic growth. This means that indicators are no longer to be only used as a discretionary policy-making tool by local planning authorities. Indeed, without a sufficient evidence base, emerging LDFs will not be declared 'sound' by Inspectors at the public examination stage and will not be able to proceed to adoption (ODPM 2004d).

Since the Public Service Agreement (PSA) 6 states that all local planning authorities have to complete their LDFs by 2006, there is a very restricted period of time available for local authorities to prepare the LDF and to compile, process and analyse indicators to develop the monitoring framework. Hence, there is a real issue of how to realistically develop robust LDF monitoring within the time-frame set out in PSA6. The University of Liverpool was commissioned by the ODPM to develop guidance on the monitoring of the LDFs (Wong et al. 2005; ODPM 2005b). The process of preparing the guidance has been similar to past practice – initiated by the centre, followed by a consultation exercise with regional and local planning bodies, and other stakeholders. A list of proposed monitoring indicators was included in the draft consultation document on Regional Spatial Strategies (ODPM 2003b), which then served as the basis for developing the monitoring indicators for the LDFs. As part of this research, inter-views, discussions and workshops involving various policy-makers and interested parties at the national, regional and local levels were carried out. The findings from this work reveal that the use of national core indicators at all spatial scales has received widespread support, although this support is conditional on the core indicators being kept to a minimum so as to allow for the development of additional local indicators. A general convergence of monitoring approaches and targets and indicators employed can, therefore, be expected as both spatial scales of monitoring (regional and local) will share: similar annual monitoring require-ments; frameworks based around objectives–targets–indicators approaches; and a common set of core indicators derived from national guidance.

Based on the discussion with local and regional policy-makers, as well as a review of literature, five broad principles were identified by the research team to underpin the monitoring framework:

1)   Making use of existing datasets.
2)   Harmonisation of indicator definitions with other existing policy and perfor-
     mance indicators.
3)   Synchronisation with Planning Policy Statements and the Regional Spatial
     Strategy monitoring framework to provide a logical and consistent approach
     to the monitoring of spatial planning across England.
4)   Adoption of the integrated 'objectives–targets–indicators' approach as the
     monitoring framework.
5)   Implementation of an embedded, analytical, forward-looking approach to
     monitoring.

A framework of indicators was proposed to monitor the LDF. Different types of
indicators play a specific role in different stages of the plan-making process.
Contextual indicators aim to enhance understanding of the wider context for the
development of spatial policies. Core national and local output indicators and
housing trajectories inform policy progress and achievement. In order to avoid
the burden of reinventing a different framework, the recommendations to local
planning authorities on the choice of contextual indicators on demographic structure, socio-cultural issues, economy, environment, housing and built environment,
and transport and spatial connectivity are closely followed by the key components used for monitoring sustainable communities in the Egan Review. Output
indicators involve the measurement of quantifiable physical outputs that demonstrate the direct effects of policy (e.g. number of housing completions, amount of
employment floorspace, etc). The national core indicators are used to reflect key
national planning priorities, and all authorities will need to assess policy implementation against these and report upon them in their annual monitoring reports.
In addition, local output indicators should be developed by local authorities to
reflect particular concerns or objectives in terms of policy performance.

     During the process of selecting the indicators, many rounds of negotiation
and discussion were made between the ODPM and representatives of local
planning authorities and regional planning bodies, as well as within the ODPM.
There has been iterative discussion to harmonise the proposed core output indicators used to monitor Regional Spatial Strategies and LDFs. While it is easier
to establish the basic principles of the monitoring and indicators framework, it is
the exact definition of each individual indicator that has caused problems. Even
when the same indicator is used, there can still be some differences in definition,
interpretation and data collection practices. More importantly, it is difficult to
find robust indicators that can gauge the direct policy output of spatial planning
policies.

     The need to harmonise different sets of indicators also runs horizontally with
other strategies, most importantly the sustainability appraisal of Local Development

Documents and the monitoring of community strategies. As part of developing their approach to LDF monitoring, authorities will need to consider how this might link with strategic environmental assessment requirements as set out in European Union Directive 2001/42/EC. Since the government has developed an integrated approach in terms of incorporating the requirements of the EU Directive within sustainability appraisal, monitoring linkages with sustainability appraisal have to be considered. As the LDFs and community strategies share the same objective of sustainable development, there should be some degree of communality in their intelligence and monitoring requirements. The advice from the ODPM is that both should adopt common targets and indicators where possible and appropriate.

As an observer of the process, one has to say that officers in the ODPM have been making concerted effort to explore the possibilities of simplifying the requirements of compiling different indicators for policy monitoring. This has been a very positive move, though the task of dealing with very complex policy issues within the complex web of political reality means that it is difficult to deliver what is desired and there is the danger of achieving a 'lowest common denominator' solution.

TENSIONS AND DILEMMAS

The process of identifying and developing sustainable and spatial planning indicators illustrates the tensions and dilemmas faced by decision-makers. The empowerment of local communities in decision-making tends to be matched by a concomitant top-down process of central guidance and the specification of core or headline indicators through the harmonisation process. This is because the global and the local have to meet their respective policy needs by collecting policy intelligence via a particular approach. It is clear that the global–local nexus of interaction is fuelled with political sensitivity; in spite of the claim that bottom-up representation and consultation is made, by and large the indicator frameworks and conceptual definitions are led by those in the upper tier of governance. In spite of the effort put in to reconcile the technical and political issues involved in developing these indicators, they are wicked problems tangled up in the web of politics. It is also clear that the practice across the Atlantic is rather different. While it is possible to find successful and genuine community indicator projects in the USA, the situation in Britain is very much that of the development of indicators for the community, implemented by local government under strong central guidance.

CHAPTER 11

CONCLUSION

> While the uses and potentialities of social indicators are not as great or immediate as originally proposed, even the more realistic and circumscribed set of applications that remain are very important and fully worthy of active pursuit.
>
> (Smith 1980/1: 742)

The central tenet of this book is to examine the interplay of instrumental rationality and normative policy context in the process of indicator development. Hence, the first part of the book sets out the wider policy context, and the political and managerial issues, which impinge on the development of indicators and the supporting statistical infrastructure; and the second part focuses on examining the latest debates over methodological issues concerning the construction and analysis of quantitative indicators. It is then, through the case study analysis of three broad groups of indicators, that some of the technical and political issues are explored in greater depth to bring out the political-managerial perspective on the technical process of indicator development. The purpose of this final chapter is, therefore, to provide a synthesis of the key issues discussed in the earlier chapters and to reflect upon the progress made today in the development and usage of indicators. Although this is largely based on detailed observations in Britain, the discussion does draw upon experience elsewhere. Finally, the discussion will be ended by looking forward at the prospect for the future development of indicator research and usage.

Through the in-depth analysis of the case study indicators, a number of key issues concerning the interplay between policy and methods in the process of indicator development can be identified. These can be consolidated into three main areas of discussion. First of all, the case studies clearly demonstrate that the methodological and conceptual development of indicators is very much influenced by the emerging policy agenda and shaped by the institutional-managerial culture of the time; and vice versa. Second, the spatial decentralisation of policy-making power is only found to be tightly controlled by a centralised approach towards policy monitoring and evaluation. Finally, the methodological and technical development of indicators very much impinges on the concomitant political forces operating in the wider context, which can facilitate as well as undermine their development. There is often a trade-off between the move towards more sophisticated methodology and the process of bottom-up consensus building. These issues will be discussed in turn to shed light on the debate over the policy practice of indicator usage and development.

Policy Ideology and Indicator Usage

One of the most obvious messages to emerge from the discussion is the increasing emphasis placed by the British government on the use of all sorts of indicators to gauge policy performance and outcomes across the full spectrum of government activities. There is a clear transition from the 1980s, when indicators were used instrumentally to allocate resources for various urban regeneration, regional assistance and housing programmes, to the current evidence-based approach toward policy-making across all sectors. The twists and turns in political ideology and government ethos over public expenditure and policy monitoring have shaped both the methodology and the usage of indicators over the last two decades. The extension of indicators in policy performance monitoring can be seen as a shift from a resource-led to a demand-led management regime. Since resources for public service delivery are highly constrained, the idea is to use consumer demand as a lever to manage the supply of these services. More importantly, the advocate of an evidence-based policy regime very much reflects the rationalist approach of policy management from the Treasury.

Far back in the 1970s, Rose (1972) argued that the more immediate the problem, the greater the willingness of policy-makers to consume information and the greater the premium upon the speed with which information can be obtained. This is still evident in current policy circles. In order to fulfil the pledge in the election manifesto for the first term of the Labour government, a Social Exclusion Unit was established to take responsibility for tackling the issue of deprivation and exclusion. Hence, the Task Team produced the *Better Information* report, which provided a critical evaluation of the problem encountered in compiling socio-economic statistics at the neighbourhood level. More recently, following the devolution of regional economic policies to the Regional Development Agencies, Christopher Allsopp was asked to examine the information base supporting economic policy development at the regional level. The Allsopp Review is particularly critical over the basis of regional productivity data, as this is the very information required by the Treasury and the Department of Trade and Industry to monitor the Tier 2 performance targets of the Regional Development Agencies. Following the newly introduced planning reforms in autumn 2004, the latest information hunt is related to the collection of planning policy indicators for the monitoring of spatial planning strategies at both regional and local levels. Similar observations were found at the European level: the ESPON programme was established to support the development of the European Spatial Development Perspective, and Eurostat has been heavily engaged in developing different indicator sets to inform European Commission policies and to monitor the EU Structural Assistance. It is these pressing policy needs that throw indicators and information needs in the political limelight, which in turn leads to the improvement of the data infrastructure and management framework. However, the relationship is not always one-way. For instance, the development of area-based deprivation indicators has triggered debate over the intrinsic value of area-based and people-based regeneration policies in tackling the problem of deprivation.

The competing demands of information and the premium of political pressure have also imposed strains on the statistical infrastructure at all spatial levels. For instance, the Allsopp Review openly comments on the tension between prioritising resources to develop neighbourhood statistics and the urgent need of improving regional productivity data. The review also comments that the demand from the European Union has created uncertainty over the national statistical infrastructure, as the Office for National Statistics has to respond to the new policy-monitoring requirements of the EU. This means that the direction and priority of data infrastructure development becomes a contested area that is increasingly subject to the politicisation of competing policy needs and biddings.

CENTRALISED DECENTRALISATION AND GLOCALISATION

The second trend observed in Britain is the concomitant centralisation of monitoring guidance and decentralisation of data collection responsibility. This has been found in the development of sustainability indicators, deprivation indicators and, more recently, performance indicators in respect of regional economic policies and spatial planning. While the need to develop indicators tailored to local circumstances is emphasised, there is strong advice and guidance over how to do it from the government and this is often accompanied by a recommended set of core indicators. This means that strong central guidance, coupled with local flexibility in developing a two-tiered indicator structure, has characterised the new indicators policy regime. It is also interesting to note that the central state is also responsible for reviewing and consolidating different recommended headline/core indicator sets, though with extensive consultation with local and regional stakeholders. However, there seems to be a missing link somewhere between such central–local arrangements. This is epitomised by the criticisms made by the Allsopp Review that, without the support of reliable regional statistics, policy-making responsibility should not be devolved. More importantly, the review argues that the quality framework of National Statistics should be extended to regional statistics; this echoes the plea made by the Royal Statistical Society when the revised framework of National Statistics was introduced in 1999.

The concomitant centralised–decentralised approaches towards indicators development have also been observed at the international level. Both the United Nations and the European Union have been very keen to encourage local communities to develop sustainability indicators to achieve the visions laid out in Agenda 21. However, they also acknowledge the problems of this hundred flowers blossoming approach, which means that it is impossible to grasp the aggregate progress made, nor to make meaningful spatial comparisons. Hence, significant effort has been made by these global institutions to devise methodologies and core indicator sets to provide a common norm. Again, significant bottom-up consultation and representation exercises were carried out in the indicator harmonisation process. This interplay of global and local politics in the process of developing indicators shows that the technical rationality of indicators is deeply entwined with the multiple webs of politics. Whilst there is strong communication with the bottom end, the process of indicator rationalisation is very much a top-down structure led by the highest level of governance.

The case of sustainability indicators clearly shows that the empowerment of decision-making power to the community and the lower spatial-tier governance tends to be quietly crawled back through the centrally designed monitoring regime. This illustrates the constant jostling of power at different spatial tiers of governance, the operation of centralised decentralisation of policy-making power, as well as the interplay of global and local politics; these political issues should thus not be neglected in the process of indicators development.

TECHNICAL EXCELLENCE AND DEMOCRACY

The third area of changing practice in indicator development has been a gradual shift in policy monitoring from a pure emphasis of outputs and implementation to the wider concern of impacts and strategic outcomes. Criticism has been made in the past by academics over the practice of the Thatcher government and the European Union in counting output indicators such as the number of jobs created and the private–public investment leverage ratios in their policy evaluation. This simple accountancy approach to policy evaluation neglects the wider impact and outcomes that regeneration policies have on society. Following the introduction of evidence-based ideology in policy-making, an important signal projected from the government's monitoring guidelines is the increasing emphasis on the longer-term horizon of outcome and impact measurement. This also coincides with the latest guidelines issued by the European Commission over the monitoring of the new programme of structural assistance. This move towards a more long-term, strategic approach to policy monitoring is encouraging.

A framework for different categories of indicators is increasingly recommended as the approach to evaluate different aspects of policy performance in the monitoring guidance documents issued by the EU and different British government departments. This includes the use of contextual indicators to establish baselines to provide a backdrop to the analysis of social change to ascertain policy outcomes. However, there is concern that practitioners on the ground may find the distinction between terminologies such as context, inputs, outputs, outcomes and impacts difficult to grasp. In some cases, such as the development of Regional Competitiveness Indicators and the Sustainability Count Indicators, the architects themselves seemed to be somewhat confused about the nature and purpose of the indicators; however, it is more likely that they were subject to all sorts of political pressure to water down the original conceptual framework by taking into account the 'bottom-up' representations.

With the pressing political pressure to fine-tune the allocation formula of the official index of deprivation, the construction of deprivation indicators has moved away from a simple, transparent approach in favour of more sophisticated statistical modelling methods, and the number of indicators used has also been largely increased. This is exemplified by both the 2000 and 2004 Indices of Multiple Deprivation. Similarly, both the United Nations and the OECD have opted to adopt a large number of sustainable development indicators based on the complex Statement–Pressure–Response causal chain model. Although the British government also has nearly 150 sustainable indicators, the Working Group decided that the causal chain model was too complicated to implement. The different approaches used in dealing with deprivation and sustainability indicators by the British government is probably due to the fact that the former are aimed for formulaic funding allocation where a seemingly impartial approach is needed, while the latter are used to inform community development at the local level and there is no pressing need to create a composite index, hence the emphasis tends to be placed on interpretation and policy enlightenment. With the constant bombardment of all sorts of monitoring indicators over the last few years, the sentiment is to rationalise these indicator sets and keep the methodology simple and transparent for wider understanding. Hence, tiered indicator structures emerged and the emphasis is to integrate the analysis and interpretation of indicator values with policy objectives and the wider policy operation environment. As Sawicki (2002) observed, community indicator projects aiming at consensus building tended to lean on using simplistic methodologies to gain a wider spectrum of communication. However, it is interesting to find that the more successful projects, such as the Jacksonville Community Indicators Project in the USA, manage to adopt a more complex causation model in their indicator framework.

The choice between different approaches to develop indicators is a difficult one because technical methodology *par excellence* is only part of the story. It is the wider politics and policy context that help to shape the decision. It is thus a fine balancing act to develop a framework that is widely accepted by the key stakeholders as well as being methodologically robust. In order to appeal to a wide spectrum of stakeholders, there is a tendency to propose more and more indicators, as was evident in the Egan Review, which still proposed four new indicators despite the starting principle of making use of existing data sources.

The pursuit of methodological excellence of some indicators sets, through major statistical processing and manipulation exercises, could undermine the transparency of the indicator creation process and stifle debate and discussion. On the other hand, the bottom-up consensus-building process could also erode the rigour of indicator development, as mediocre solutions tend to be adopted to satisfy the interest of all stakeholders. The discussion here highlights that there are inherent tensions between complexity and transparency, and between a 'one size fits all' type of solution and the purity of technical excellence. Such problems will always exist and be a politically contested area of debate, especially when resources and performance assessments are attached to the indicators.

WHERE WILL THE FUTURE LIE?

At the beginning of the book, some big questions were posed in the light of the current trends in indicator development. Learning from the experience of the social indicators movement, there is a concern over whether indicators will be embedded into the governance structure, or just another trial and error experiment that will fade out in a few years time. Charting the history of indicator development from the 1970s to now clearly shows that indicators have had their ups and downs in the policy-making process. One interesting difference is that the social indicators movement tended to aim for grand indicators systems and national social welfare, whereas the current mood of indicators development is much more engrained with the management of community and local governance. It may be somewhat too early to predict whether indicators and evidence-based policy-making will go through another boom and bust cycle; the signs so far show that the use of indicators has been cautious rather than fuelled with optimism, and stakeholders have a greater appreciation and realism of the extent to which indicators can inform decision-making. This cautious attitude may be particularly justified in the case of Britain as the enthusiasm towards Local Agenda 21 did not last long and has been replaced by the centrally directed community strategies with a recommended set of indicators from the centre. The paradox is that a slow warming-up process of embracing indicators into the decision-making process may be a more sustainable one. Another positive sign is that indicator development has been a technical exercise as well as a joined-working and consensus-building process.

In spite of the comment that there is no evidence to show that indicators or research generally influence the ultimate policy decisions, politicians still prefer to lean on data and statistics to inform their judgements, as illustrated by the ESPON programme. This is partly because indicators provide a systematic, uniform approach to examine issues of concern and track progress made; and partly because there are no other more attractive alternatives as decisions have to be made on some basis. The discussion in the book illustrates that the interface between technical and normative rationality makes indicators attractive to policy-makers because the concepts to be measured can be shifted, and the indicators used can be adjusted. There is thus a certain degree of flexibility in the deployment of indicators that is very suitable in coping with the continuously changing problem definitions in policy discourses.

Another way to look at the issue is that, in spite of the swinging pendulum, indicators have never totally vanished from the policy arena. The latest comeback of indicators may reinforce the belief that there are some intrinsic values and underlying strengths to indicators that they merit inclusion as part of the policy-making instruments. Based on the experience in Britain, it is not difficult to pinpoint the progress made over the last two decades. There have been incremental developments to fine-tune the methodology of indicator construction. These include using modelling methods to estimate indicator values for small areas; exploring different methods to combine indicators; using different approaches to present and analyse indicator values; and coming to an acceptance that not all issues can be quantified and measured. In spite of the fact that many indicator sets still follow an empirical data-driven approach, there is an expectation that there is a need to clarify and define the concept to be measured, though at times inconsistencies were found in the definition, such as the definition of the Department of Trade and Industry's Regional Competitiveness Indicators. There is also a general acceptance of the policy enlightenment function of indicators, hence measurement is only part of the job and there is a need to analyse and interpret the indicators. Stronger emphasis has also been placed on user engagement in the development and harmonisation process of indicator sets, which shows that both the rational paradigm and the communicative, social learning approach towards indicator development are operating in parallel. Through the pressing need for more indicators, there is a higher degree of political awareness of the inadequacy of data infrastructure at all spatial levels and the urgent need to connect policy requirements with data availability. All these have led to more joined-up thinking and cross-departmental work towards the development of indicator sets such as quality of life indicators.

While social indicators are seen as serving an important role in contemporary democratic society, and some progress has been made, it is clear that there are plenty of challenges ahead. First of all, there has to be a realisation that the technical dimension of indicator research is tightly constrained by the wider political and institutional process of data management and organisation at the highest level. For instance, resources have to be allocated to improve the data infrastructure to support policy needs at all spatial levels. Second, since indicators do not sit comfortably between the stools of rationalist and empiricist ideology, there is an inherent tension in the reconciliation of different sets of values, especially on how to secure objective knowledge from belief, opinion and even prejudice that is somewhat less than convincing. There is also a need to reduce the tension between top-down and bottom-up approaches to indicator development: for example, should the stewardship role of the higher-level governance in overseeing local indicators development be more explicit? Third, as many policy problems are wicked problems and open concepts, the use of indicators to monitor policy outcomes requires a clear agreement of what needs to be improved, and what does improvement entail. Hence the government statistician, Christopher Day, wisely commented that 'It is easy enough to criticize the indicators and it would be foolish to discount these criticisms, and anyone making decisions on the evidence of indicators alone is counting disaster' (Day 1989: 2).

In conclusion, developing indicators is a very complex process, and reality often falls short of ideal when confronting problems such as data availability constraints and political pressures. The motto of indicator analysis should, therefore, be one of understanding the methodological limitations and not overexaggerating the findings. After all, indicators are not 'exact science'; they only 'indicate' and provide a useful lens to identify and highlight interesting patterns of development that merit further analysis and exploration. Also, there needs to be a realisation that indicators are not a panacea to policy problems; there are certain issues that are not susceptible to measurement and hence this approach should not be pushed too far. The famous quote from Laurence J. Peter provides some sensible advice, 'Some problems are so complex that you have to be highly intelligent and well informed just to be undecided about them' (Laurence J. Peter, famous educator and writer). Whilst indicators offer handles for us to ascertain the nature of socio-economic change and progress in policy implementation, it is only one form of knowledge, and one that has to be validated by other professional experience and knowledge. As Emile Durkheim commented nearly a century ago:

> We can no longer accept a single, invariable system of categories or intellectual frameworks; what is required is the tolerance of the intellectual, of the scientist, who knows that truth is a complex thing and understands that there is an excellent chance that no one of us will see the whole of its aspects.
>
> (Durkheim 1913–14: 71, quoted in Bryant 1985: 43)

REFERENCES

AIU [Adding It Up] (2003) 'Policy choice', 4 February 2003, Adding It Up website (www.addingitup.gov.uk/epc/epc _ overview1.cfm).

Allsopp, C. (2004) *Review of Statistics for Economic Policymaking*, final report to the Chancellor of the Exchequer, the Governor of the Bank of England and the National Statistician, Norwich: HMSO.

Altvater, E. (1992) 'Fordist and post-Fordist international division of labour and monetary regimes', in M. Storper and A.J. Scott (eds) *Pathways to Industrialization and Regional Development*, London: Routledge.

Ambiente Italia (2003) *European Common Indicators: Development, Refinement, Management and Evaluation*, final report to the European Commission, Milan: Italy, Ambiente Italia Research Institute.

Ameritrust Corporation and SRI International (1986) *Indicators of Economic Capacity*, Menlo Park, California: SRI International.

Amin, A. and Thrift, N. (1994) *Globalization, Institutions, and Regional Development in Europe*, Oxford: Oxford University Press.

Anderson, C. (1998) 'Discussion at the meeting on "alternatives to economic statistics as indicators of national well-being"', *Journal of Royal Statistical Society A* 161: 304–5.

Archibugi, F. (1998) 'Measuring urban life quality: some methodological warnings' in proceedings of the First International Conference on Quality of Life in Cities, 4–6 March, University of Singapore, Volume 3: 23–40.

ARO [Association of Regional Observatories] (2003) Information from the website (www.regionalobservatories.org.uk).

ARO [Association of Regional Observatories] (2004a) Website (www.regionalobservatories.org.uk).

ARO [Association of Regional Observatories] (2004b) 'Association of Regional Observatories Business Plan: 2004–2007, ARO, Room 606, City House, New Station Street, Leeds LS1 4US', website (www.regionalobservatories.org.uk/aro _ business _ plan.pdf).

ARO [Association of Regional Observatories] (2004c) *The State of Regional Intelligence*, Plymouth: Association of Regional Observatories.

Arup Economics and Planning (1995) *Environmental Capacity: A Methodology for Historic Cities*, a final report to Cheshire County Council, Chester City Council, Department of the Environment and English Heritage, London: Arup Economics and Planning.

Atkinson, G. (1998) 'Discussion at the meeting on "alternatives to economic statistics as indicators of national well-being"', *Journal of Royal Statistical Society A* 161: 307.

Audit Commission (1989) *Urban Regeneration and Economic Development: The Local Government Dimension*, London: HMSO.

Audit Commission (1989) (2000) *On Target: The Practice of Performance Indicators*, London: Audit Commission for Local Authorities and the National Health Service in England and Wales.

Audit Commission (1989) (2002) *Quality of Life: Using Quality of Life Indicators*, London: Audit Commission.

Babbie, E. (1992) *The Practice of Social Research*, 6th edn, Belmont, California: Wadsworth.

Bailey, N., Docherty, I. and Turok, I. (2002) 'Dimensions of city competitiveness: Edinburgh and Glasgow in a UK context', in I. Begg (ed.) *Urban Competitiveness: Policies for Dynamic Cities*, Bristol: Policy Press.

Bailey, N., Flint, J., Goodlad, R., Shucksmith, M., Fitzpatrick, S. and Pryce, G. (2003) *Measuring Deprivation in Scotland: Developing a Long-Term Strategy*, an interim report, Glasgow: Scottish Centre for Research on Social Justice.

Bains Committee (1972) *The New Local Authorities: Management and Structure*, London: HMSO.

Baker, M. and Wong, C. (1997) 'Planning for housing land in the English regions: a critique of household projections and Regional Planning Guidance mechanisms', *Environment and Planning C: Government and Policy* 15: 73–87.

Baker, M. and Wong, C. (2001) 'The use of regional intelligence to inform regional policy', paper presented at the Regional Studies Association Annual Conference 'Regionalising the Knowledge Economy', 21 November, London, Conference Proceedings: 74–7.

Baker, M. and Wong, C. (2004) 'Developing regional intelligence capacities in the English Regions', paper presented at the Association of European Planning School 2004 Congress, 1–4 July, Grenoble, France.

Bartholomew, D.J. (1988) *Measuring Social Disadvantage and Additional Educational Needs*, a report to the Department of Environment, London: Department of Statistics and Mathematical Sciences, London School of Economics.

Bartholomew, D.J. (1995) 'The measurement of unemployment in the UK (with discussion)', *Journal of Royal Statistical Society A* 158: 363–417.

Batey, P. and Brown, B. (1995) 'From human ecology to customer targeting: the evolution of geodemographics', in P. Longley and G. Clarke (eds) *GIS for Business and Service Planning*, Cambridge: GeoInformation International.

Bauer, R.A. (1966) *Social Indicators*, Cambridge, MA: MIT.

Beatty, C., Fothergill, S., Gore, T. and Herrington, A. (1997) *The Real Level of Unemployment*, Centre for Regional Economic and Social Research, Sheffield: Sheffield Hallam University.

Beauregard, R.A. (1993) 'Constituting economic development: a theoretical perspective', in R.D. Bingham and R. Mier (eds) *Theories of Local Economic Development*, California: Sage.

Becker, R.A., Denby, L., McGill, R. and Wilks, A.R. (1987) 'Analysis of data from the *Places Rated Almanac*', *The American Statistician* 41(3): 169–86.

Begg, I.G. and Cameron, G.C. (1988) 'High technology location and the urban areas of Greater Britain', *Urban Studies* 25: 361–79.

Begg, I., Moore, B. and Altunbas, Y. (2002) 'Long-run trends in the competitiveness of British cities', in I. Begg (ed.) *Urban Competitiveness: Policies for Dynamic Cities*, Bristol: Policy Press.

Bell, D. (1990) *Data Sources for Area Prioritisation: Section A – Review and Analytical Topics*, Edinburgh: DG Information Services.

Bell, S. and Morse, S. (1999) *Sustainability Indicators*, London: Earthscan.

Bell, T.L. (1984) 'The *Places Rated Almanac*: flawed out but pedagogically useful', *Journal of Geography* 82: 285–90.

Bennett, R.J. and Krebs, G. (1991) *Local Economic Development: Public–Private Partnership Initiative in Britain and Germany*, London: Belhaven Press.

Berger, P.L. and Kellner, H. (1982) *Sociology Reinterpreted: An Essay on Method and Vocation*, Harmondsworth: Pelican Books.

Berthoud, R. (1983) 'Who suffers social disadvantage?' in M. Brown (ed.) *The Structure of Disadvantage*, SSRC/DHSS Studies in Deprivation and Disadvantage 12, London: Heinemann Educational Books.

Betson, D.M., Citro, C.F. and Michael, R.T. (2000) 'Recent developments for poverty measurement in US official statistics', *Journal of Official Statistics* 16(2): 87–111.

Biehl, D. (1986) *The Contribution of Infrastructure to Regional Development*, Luxembourg: Commission of the European Communities.

Blackman, T. (1995) *Urban Policy in Practice*, London: Routledge.

Blackman, T. (1998) 'Towards evidence-based local government: theory and practice', *Local Government Studies* 24(2): 56–70.

Blakely, E.J. (1994) *Planning Local Economic Development: Theory and Practice*, 2nd edn, California: Sage.

Blowers, A. and Young, S. (2000) 'Britain: unstainable cities', in N. Low, B. Gleeson, I. Elander and R. Lidskog (eds) *Consuming Cities: The Urban Environment in the Global Economy after the Rio Declaration*, London: Routledge.

Boddy, M. (2002) 'Linking competitiveness and cohesion', in I. Begg (ed.) *Urban Competitiveness: Policies for Dynamic Cities*, Bristol: Policy Press.

Boddy, M. and Snape, D. (1995) *The Role of Research in Local Government*, a final report to the Local Authorities Research and Intelligence Association (LARIA), Wokingham: LARIA.

Bosman, J. and de Smidt, M. (1993) 'The geographical formation of international management centres in Europe', *Urban Studies* 30(6): 967–80.

Bovaird, T. (1995) 'Urban governance and quality of life marketing – emerging contradictions in strategies for urban competition', paper presented at Regional Studies Association European Conference, Gothenburg, Sweden, 6–9 May.

Boyer, R. and Savageau, D. (1981) *Places Rated Almanac*, Chicago: Rand McNally.

Boyer, R. and Savageau, D. (1983) *Places Rated Almanac*, Chicago: Rand McNally.

Boyer, R. and Savageau, D. (1985) *Places Rated Almanac*, Chicago: Rand McNally.

Boyle, M.R. (1989) 'Grading the state business climate report cards', *The Economic and Demographic Trends Newsletter* 2(1): 1–6.

Breheny, M. (ed.) (1999) *The People: Where Will They Work?* London: Town and Country Planning Association.

Breheny, M. and Hall, P. (eds) (1996) *The People: Where Will They Go?* London: Town and Country Planning Association.

Breheny, M.; Hall, P. and Hart, D. (1987) *Northern Lights: A Development Agenda for the North in the 1990s,* Preston: Derrick Wade & Waters.

Briggs, D., Kerrell, E., Stansfield, M. and Tantrum, D. (1995) *State of the Countryside Environment Indicators*, a final report to the Countryside Commission, Northampton: Nene Centre for Research.

Brown, M. (ed.) (1983) *The Structure of Disadvantage*, SSRC/DHSS Studies in Deprivation and Disadvantage 12, London: Heinemann Educational Books.

Brown, P.J.B. (1989) 'A super profile based affluence ranking of OPCS urban areas', *Built Environment* 14(2): 118–34.

Brown, P.J.B. and Batey, P.W.J. (1994) *Characteristics of Super Profile Lifestyles and Target Markets: Index Tables, Pen Pictures and Geographical Distribution*, The Urban Research and Policy Evaluation Regional Research Laboratory Super Profile Technical Note 2, Liverpool: University of Liverpool.

Bryant, G.A. (1985) *Positivism in Social Theory and Research*, Basingstoke: Macmillan.

Bulmer, M. (ed.) (1977) *Sociological Research Methods: An Introduction*, London: Macmillan.

Burton, P. and Boddy, M. (1995) 'The changing context for British urban policy', in R. Hambleton and H. Thomas (eds) *Urban Policy Evaluation: Challenge and Change*, London: Paul Chapman.

Cameron, G.C. (1990) 'First steps in urban policy evaluation in the United Kingdom', *Urban Studies* 27(4): 475–95.

Camp, R. (1989) *Benchmarking: The Search for Industry Best Practices that Lead to Superior Performance*, Milwaukee: Quality Press, American Society for Quality Control.

Cannell, M.G.R., Palutikof, J.P. and Sparks, T.H. (1999) *Indicators of Climate Change in the UK*, London: DETR.

Caplan, N. and Barton, E. (1978) 'The potential of social indicators: minimum conditions for impact at the national level as suggested by a study of the use of *Social Indicators 73*', *Social Indicators Research* 5: 427–56.

Carley, M. (1980) *Rational Techniques in Policy Analysis*, London: Heinemann Educational Books.

Carley, M. (1981) *Social Measurement and Social Indicators*, London: George Allen & Unwin.

Carlisle, E. (1972) 'The conceptual structure of social indicators', in A. Shonfield and S. Shaw (eds) *Social Indicators and Social Policy*, published for the Social Science Research Council, London: Heinemann Educational Books.

Carmines, E.G. and Zeller, R.A. (1979) *Reliability and Validity Assessment*, Sage University Paper Series on Quantitative Applications in the Social Sciences, 07–017, Sage: Beverley Hills.

Carstairs, V. and Morris, R. (1989) 'Deprivation: explaining difference in mortality between Scotland and England and Wales', *British Medical Journal* 299: 886–9.

Castells, M. (1989) *The Information City*, Oxford: Blackwell.

Castells, M. and Hall, P. (1994) *Technopoles of the World*, London: Routledge.

Cazes, B. (1972) 'The development of social indicators: a survey', in A. Shonfield and S. Shaw (eds) *Social Indicators and Social Policy*, published for the Social Science Research Council, London: Heinemann Educational Books.

CBI [Confederation of British Industry] (1997) *Regions for Business: Improving Policy Design and Delivery*, London: Confederation of British Industry.

CEC [Commission of the European Communities] (1991) *Annual Report on the Implementation of the Reform of the Structural Funds 1989*, Luxembourg: Office for Official Publications of the European Communities.

CFED [Corporation for Enterprise Development] (1991) *The 1991 Development Report Card for the States: A Tool for Public and Private Sector Decision Makers*, Washington: Corporation for Enterprise Development.

Chalmers, C. (2001) 'Combining domain indices into a general index – a valid process', London: Royal Statistical Society, website (http://stats.lse.ac.uk/galbrait/indices/).

Champion, A., Atkins, D., Coombes, M. and Fotheringham, S. (1998) *Urban Exodus*, London: Council for the Protection of Rural England.

Champion, A. and Green, A. (1990) *The Spread of Prosperity and North–South Divide*, Gorsforth and Kenihorth: Booming Towns.

Cheshire, P. (1995) 'Territorial competition: some lessons for policy', paper presented at the Regional Studies Association European Conference: Regional Futures: Past and Present, East and West, Gothenburg, Sweden, 6–9 May.

Cheshire, P., Carbonaro, G. and Hay, D. (1986) 'Problems of urban decline and growth in EEC countries: or measuring degrees of elephantness', *Urban Studies* 2: 131–49.

Citro, C.F. and Michael, R.T. (eds) (1995) *Measuring Poverty: A New Approach*, Panel on Poverty and Family Assistance: Concepts, Information Needs, and Measurement Methods, Committee on National Statistics, National Research Council, Washington, DC: National Academy Press.

Clark, T.N. (1973) 'Community social indicators: from analytical models to policy applications', *Urban Affairs Quarterly* 9(1): 3–36.

Comte, A. (1844) *A Discourse on the Positive Spirit*, prefaced and trans. E.S. Beesly, 1903, London: William Reeves.

Connolly, C. and Chisholm, M. (1999) 'The use of indicators for targeting public expenditure: the Index of Local Deprivation', *Environment and Planning C: Government and Policy* 17: 463–82.

Cook, L. and Martin, J. (2005) '35 years of social change'. *Social Trends*, 35: 1–6.

Cooke, P. and Morgan, K. (1993) 'The network paradigm', *Environment and Planning D* 11: 543–64.

Coombes, M. and Raybould, S. (1989) 'Developing a Local Enterprise Activity Potential (LEAP) index', *Built Environment* 14: 107–17.

Coombes, M., Raybould, S. and Wong, C. (1992) *Developing Indicators to Assess the Potential for Urban Regeneration*, London: HMSO.

Coombes, M., Raybould, S., Wong, C. and Openshaw, S. (1995) *The 1991 Deprivation Index: A Review of Approaches*, London: HMSO.

Coombes, M. and Wong, C. (1994) 'Methodological steps in the development of multi-variate indexes for urban and regional policy analysis', *Environment and Planning A* 26: 1297–316.

Coombes, M., Wong, C. and Raybould, S. (1993a) *Indicators of Disadvantage and the Selection of Areas for Regeneration*, Edinburgh: Scottish Homes Research Report No. 26.

Coombes, M., Wong, C. and Raybould, S. (1993b) *Local Environment Index: Infrastructural Resources Dimension*, a final report to the Department of Employment, Centre for Urban and Regional Development Studies, Newcastle upon Tyne: University of Newcastle upon Tyne.

Copus, A.K. and Crabtree, J.R. (1996) 'Indicators of socio-economic sustainability: an application to remote rural Scotland', *Journal of Rural Studies* 12(1): 41–54.

Countryside Agency (2000) 'New index – new opportunities for rural areas', 23 August 2000, website (www.countryside.gov.uk/WhoWeAreAndWhatWeDo/pressCentre/new _ index.asp).

Countryside Agency (2001) *Indicators of Rural Disadvantage: The North East*, a report prepared for consultation by Corporate Planning, Research, Data and Information, Cheltenham: Countryside Agency.

Countryside Commission, English Heritage, English Nature (1993) *Conservation Issues in Strategic Plans*, Cheltenham: Countryside Commission.

Cullingworth, B. and Nadin, V. (2002) *Town and Country Planning in the UK*, 13th edn, London: Routledge.

CURDS [Centre for Urban and Regional Development Studies] (1988) *Area Economic Development Studies – Kirklees, Wakefield and Doncaster Co-ordination Report*, London: HMSO.

Curnow, R. (1998) 'Editorial: an independent national statistical services', *Journal of Royal Statistical Society A* 161: 275–7.

Custance, J. and Hillier, H. (1998) 'Statistical issues in developing indicators of sustainable development', *Journal of Royal Statistical Society A* 161, Part 3: 281–90.

Davies, J. (1996) 'The politics of experience: urgent lessons on safeguarding data as local authorities reorganise again', *Mapping Awareness* 10(4): 28–31.

Day, C. (1989) *Taking Action with Indicators*, London: HMSO.

Deakin, N. (1982) 'Research and the policy-making process in local government', *Policy and Politics* 10(3): 303–15.

Deas, I., Robson, B., Wong, C. and Bradford, M. (2003) 'Measuring neighbourhood deprivation: a critique of the Index of Multiple Deprivation', *Environment and Planning C* 21(6): 883–903.

Debbage, K. and Ree, J. (1991) 'Company perceptions of comparative advantage by region', *Regional Studies* 25(3): 199–206.

DEFRA [Department of the Environment, Food and Rural Affairs] (2004) DEFRA sustainability indicators web page (www.sustainable-development.gov.uk/indicators/index.htm).

DEFRA [Department of the Environment, Food and Rural Affairs] (2004) *UK Government Quality of Life Counts: Update 2004*, London: DEFRA.

De Neufville, J.I. (1975) *Social Indicators and Public Policy: Interactive Processes of Design and Application*, Amsterdam: Elsevier Publishing.

DETR [Department of the Environment, Transport and the Regions] (1997a) *Supplementary Guidance for Round 4 of the Single Regeneration Budget Challenge Fund*, London: Department of the Environment.

DETR [Department of the Environment, Transport and the Regions] (1997b) *Regeneration Programmes: The Way Forward*, London: Department of the Environment.

DETR [Department of the Environment, Transport and the Regions] (1997c) *Building Partnerships for Prosperity: Sustainable Growth, Competitiveness and Employment in the English Regions*, Cm 3814, London: Stationery Office.

DETR [Department of the Environment, Transport and the Regions] (1998a) *Sustainability Counts*, London: Department of the Environment, Transport and the Regions.

DETR [Department of the Environment, Transport and the Regions] (1998b) *1998 Index of Local Deprivation: A Summary of Results*, London: Department of the Environment, Transport and the Regions.

DETR [Department of the Environment, Transport and the Regions] (1998c) *The Future of Regional Planning Guidance*, London: Department of the Environment, Transport and the Regions.

DETR [Department of the Environment, Transport and the Regions] (1998d) *Planning for the Communities of the Future*, London: Stationery Office.

DETR [Department of the Environment, Transport and the Regions] (1998e) *Sustainable Development: Opportunities for Change, Consultation Paper on a Revised UK Strategy*, London: Department of the Environment, Transport and the Regions.

DETR [Department of the Environment, Transport and the Regions] (1999a) *Regional Development Agencies' Regional Strategies*, London: Department of the Environment, Transport and the Regions.

DETR [Department of the Environment, Transport and the Regions] (1999b) *Monitoring Progress: Indicators for the Strategy for Sustainable Development in the UK*, London: Department of the Environment, Transport and the Regions.

DETR [Department of the Environment, Transport and the Regions] (1999c) *Performance Indicators for 2000/2001*, a joint consultation document produced by DETR and the Audit Commission on Best Value and local authority performance indicators for 2000/2001, London: Department of the Environment, Transport and the Regions.

DETR [Department of the Environment, Transport and the Regions] (1999d) *Local Government Act 1999: Part I Best Value*, DETR Circular, October 1999, London: Department of the Environment, Transport and the Regions.

DETR [Department of the Environment, Transport and the Regions] (1999e) *Towards an Urban Renaissance*, final report of the Urban Task Force, London: Stationery Office.

DETR [Department of the Environment, Transport and the Regions] (2000a) *Local Quality of Life Counts: A Handbook for a Menu of Local Indicators of Sustainable Development*, London: Department of the Environment, Transport and the Regions.

DETR [Department of the Environment, Transport and the Regions] (2000b) *Measuring Multiple Deprivation at the Small Area Level: The Indices of Deprivation 2000*, London: Department of the Environment, Transport and the Regions.

DETR [Department of the Environment, Transport and the Regions] (2000c) *Monitoring Provision of Housing through the Planning System towards Better Practice*, London: Department of the Environment, Transport and the Regions.

DETR [Department of the Environment, Transport and the Regions] (2000d) *Our Towns and Cities: The Future (Delivering an Urban Renaissance)*, London: DETR.

DETR [Department of the Environment, Transport and the Regions] (2000e) *Allocation of Housing Capital Resources 2001/02*, London: Department of the Environment, Transport and the Regions.

Diamond, D. and Spence, N.A. (1989) *Infrastructure and Industrial Costs in British Industry*, London: HMSO.

DoE [Department of the Environment] (1983) *Urban Deprivation*, Information Note 2, Inner City Directorates, London: Department of the Environment.

DoE [Department of the Environment] (1993) *Environmental Appraisal of Development Plans*, London: HMSO.

DoE [Department of the Environment] (1994) *Sustainable Development: The UK Strategy*, London: HMSO.

DoE [Department of the Environment] (1995) *Partners in Regeneration: The Challenge Fund Bidding Guidance*, London: Department of the Environment.

DoE [Department of the Environment] (1996) *Indicators of Sustainable Development for the United Kingdom*, London: HMSO.

DoE [Department of the Environment] (1997) *Partners in Regeneration: The Challenge Fund Bidding Guidance,* London: Department of the Environment.

Doeringer, P.B., Terkla, D.G. and Topakian, G.C. (1987) *Invisible Factors in Local Economic Development*, New York: Oxford University Press.

Donnison, D. (1975) 'The age of innocence is past: some ideas about urban research and planning', *Urban Studies* 12: 263–72.

Dorling, D. (ed.) (2001) 'How much does place matter?' *Environment and Planning A* 33: 1335–69.

DTI [Department of Trade and Industry] (1997) *Regional Competitiveness Indicators*, a consultation paper, London: Department of Trade and Industry.

DTI [Department of Trade and Industry] (1998) *Regional Competitiveness Indicators*, London: Department of Trade and Industry.

DTI [Department of Trade and Industry] (1999) 'Regional development agencies: regional development agency core indicators', website (www.dti.gov.uk/ rda/info/core.htm).

DTI [Department of Trade and Industry] (2002a) *Regional Competitiveness and State of the Regions*, London: Department of Trade and Industry.

DTI [Department of Trade and Industry] (2002b) 'Tier 2 technical note 2', DTI website (www.consumer.gov.uk/rda/info/tier2tech.htm).

DTI [Department of Trade and Industry] (2003) 'Regional development agencies', DTI website (www.dti.gov.uk/rda/info).

DTI [Department of Trade and Industry] (2004) *Regional Competitiveness and State of the Regions*, London: Department of Trade and Industry.

DTI [Department of Trade and Industry] (2005) *Regional Competitiveness and State of the Regions*, London: Department of Trade and Industry.

DTI, DETR, The Welsh Office and The Scottish Office (1999) *The Government's Proposal for New Assisted Areas*, London: Department of Trade and Industry.

DTI, The Scottish Office, The Welsh Office (1993) *Regional Policy Review of the Assisted Areas of Great Britain*, London: Department of Trade and Industry.

DTLR [Department for Transport, Local Government and the Regions] (2002) *Urban Policy Evaluation Strategy Consultation Document*, London: DTRL.

Duguid, G. and Grant, R. (1983) *Areas of Special Need in Scotland*, Edinburgh: Scottish Office General Research Unit.

Duncan, O.D. (1969) *Towards Social Reporting: Next Steps*, New York: Russell Sage Foundation.

Dunford, M. (1990) 'Theories of regulation', *Environment and Planning D: Society and Space* 8: 297–321.

Dunn, J., Hodge, I., Monk, S. and Kiddle, C. (1998) *Developing Indicators of Rural Disadvantage*, a final report to the Rural Development Commission, Cambridge: Department of Land Economy, University of Cambridge.

Dunnell, K. (2002) 'Neighbourhood Statistics: concepts and frameworks', a paper on the Office for National Statistics's Neighbourhood Statistics website (www.neighbourhood.statistics.gov.uk/latest-news.asp), 12 November 2002: 1–18.

Duvall, R. and Sharmir, M. (1980) 'Indicators from errors: cross-national, time-serial measures of the repressive disposition of governments', in C.L. Taylor (ed.) *Indicator Systems for Political, Economic and Social Analysis*, Cambridge, MA: Gunn & Hain.

Edwards, J. (1975) 'Social indicators, urban deprivation and positive discrimination', *Journal of Social Policy* 4: 275–87.

Egan, J. (2004) *Skills for Sustainable Communities*, London: RIBA Enterprises Ltd.

Eisenschitz, A. and Gough, J. (1993) *The Politics of Local Economic Policy: The Problems and Possibilities of Local Initiative*, Basingstoke: Macmillan.

EMRA [East Midlands Regional Association] (2001) Information from the website (www.eastmidlandsobservatory.org.uk).

Ernst and Young (1998) *Benchmarking Regional Competitiveness*, an unpublished report for the Government Offices of the East and West Midlands, Birmingham: Ernst and Young.

ESPON [European Spatial Planning Observation Network] (2005) Official Website of the ESPON 2006 Programme, supported by the EU-Community Initiative Interreg III (www.espon.lu).

European Commission (1988) 'Sustainable urban development in the European Union: a framework for action', Commission Communication, COM 605, Brussels: European Commission.

European Commission (2000a) 'Indicators for monitoring and evaluation: an indicative methodology', The New Progamming Period 2000–2006 Methodological Working Papers 3, Brussels: European Commission.

European Commission (2000b) *The Urban Audit: Towards the Benchmarking of Quality of Life in 58 European Cities Volume 1: The Yearbook*, Luxembourg: Office for the Official Publications of the European Communities.

European Commission (2000c) *Towards a Local Sustainability Profile: European Common Indicators*, Working Group on Measuring, Monitoring and Evaluation in Local Sustainability, Brussels: European Commission.

Eurostat [Statistical Office of the European Communities] (2001a) *Environmental Pressure Indicators for the EU*, Luxembourg: Eurostat.

Eurostat [Statistical Office of the European Communities] (2001b) *Measuring Progress Towards a More Sustainable Europe*, Luxembourg: Eurostat.

Evans, A. (2003) 'Shouting very loudly: economics, planning and politics', *Town Planning Review* 74(2): 195–212.

Fenwick, J. (1992) 'Policy research in local government', *Local Government Policy Making* 18(4): 35–41.

Ferge, Z. (1987) 'Studying poverty', in Z. Ferge and S.M. Miller (eds) *Dynamics of Deprivation*, Hants: Aldershot.

Fieldhouse, E.A. and Tye, R. (1996) 'Deprived people or deprived places? Exploring the ecological fallacy in studies of deprivation with Samples of Anonymised Records', *Environment and Planning A* 28: 237–59.

Fielding, A. and Halford, S. (1990) *Patterns and Processes of Urban Change in the United Kingdom*, London: HMSO.

Findlay, A.; Rogerson, R. and Morris, A. (1989) 'In what sense "indicators" of quality of life?' *Built Environment* 14(2): 96–106.

Firth, D., Noble, M. and Smith, G. (2001) 'Indices of deprivation', correspondence, *Journal of Royal Statistical Society A* 164: 566.

Flynn, P. (1986) 'Urban deprivation: what it is and how to measure it?' *Public Money*, September 1996: 37–41.

Forrest, R. and Gordon, D. (1993) *People and Places: A 1991 Census Atlas of England*, School for Advanced Urban Studies in association with the Bristol Statistical Monitoring Unit, Bristol: School for Advanced Urban Studies.

Forrest, R. and Kearns, A. (2001) 'Social cohesion, social capital and the neighbourhood', *Urban Studies* 38(12): 2125–43.

Fox, K.A. (1974) *Social Indicators and Social Theory: Elements of an Operational System*, New York: Wiley-Interscience.

Frey, R.L. (1979) *Die Infrastruktur als mittel der Regionalpolitik*, Bern: Haupt.

Giannias, D., Liargovas, P. and Manolas, G. (1999) 'Quality of life indices for analysing convergence in the European Union', *Regional Studies* 33(1): 27–35.

Gilfoyle, I. and Wong, C. (1998) 'Computer applications in planning: twenty years' experience of Cheshire County Council', *Planning Practice and Research* 13(2): 191–7.

Gordon, D. (1995) 'Census based deprivation indices: their weighting and validation', *Journal of Epidemiology and Community Health* 49, Suppl. 2: S39–S44.

Gordon, D. and Forrest, R. (1995) *People and Places II: Social and Economic Distinctions in England*, Bristol: School of Advanced Urban Studies, University of Bristol.

Gordon, D. and Pantazis, C. (eds) (1995) *Breadline Britain in the 1990s*, report to Joseph Rowntree Foundation, York: Joseph Rowntree Foundation.

Gordon, I. (1999) 'Move on up the car: dealing with structural unemployment in London', *Local Economy* 14: 87–95.

Gottmann, J. and Harper, R.A. (1990) *Since Megalopolis: The Urban Writings of Jean Gottmann*, Baltimore: The Johns Hopkins University Press.

Gowman, N. (2000) 'The neighbourhood in focus: "this place does my head in"', the King's Fund Regeneration and Mental Health, Briefing 3, May 2000: 2.

Grayson, L. and Young, K. (1994) *Quality of Life in Cities*, London: British Library.

Green, A. (2001) 'Unemployment, nonemployment, and labour-market disadvantage', *Environment and Planning A* 33: 1361–4.

Green, A. and Champion, A. (1989) 'Measuring local economic performance: methodology and applications of the Booming Towns approach', *Built Environment* 14(2):78–95.

Green, A. and Champion, A. (1991) 'Booming towns studies: methodological issues', *Environment and Planning A*, 23: 1393–408.

Green, A., Davies, R., Elias, P., Hasluck, C., Owen, D. and Wilson, R. (2002) *Data Catalogue*, a report to the Department for Transport, Local Government and the Regions, Warwick: Institute for Employment Research, University of Warwick.

Green, A., Davies, R., Elias, P., Hasluck, C., Owen, D. and Wilson, R. (1991) 'Booming towns studies: methodological issues', *Environment and Planning A* 23: 1393–408.

Green, A. and Owen, D. (2002) *Regional and Sub-regional Information/Intelligence: Needs, Uses, Gaps and Priorities*, a report to the Department for Transport, Local Government and the Regions, Warwick: Institute for Employment Research, University of Warwick.

Greer, S. (1969) *The Logic of Social Inquiry*, Chicago: Aldine.

Gurr, T.R. (1981) 'A conceptual system of political indicators', in C.L. Taylor (ed.) *Indicator Systems for Political, Economic and Social Analysis*, Cambridge, MA: Gunn & Hain.

Hall, P. (1983) 'The Anglo-American connection: rival rationalities in planning theory and practice, 1955–1980', *Environment and Planning B: Planning and Design* 10: 41–46.

Hall, P., Breheny, M., McQuaid, R. and Hart, D. (1987) *Western Sunrise: The Genesis and Growth of Britain's Major High Tech Corridor*, Hemel Hempstead: Allen & Unwin.

Hambleton, R. and Thomas, H. (1995) 'Urban policy evaluation – the contours of the debate', in R. Hambleton and H. Thomas (eds) *Urban Policy Evaluation: Challenge and Change*, London: Paul Chapman Publishing.

Hammond, A., Adriaanse, A., Rodenburg, E., Bryant, D. and Woodward, R. (1996) *Environmental Indicators: A Systematic Approach to Measuring and Reporting on Environmental Policy Performance in the Context of Sustainable Development*, Washington, DC: World Resource Institute.

Haringey Council (2004) 'Index of Multiple Deprivation 2004 (Revised)', Haringey Council web page (http://haringey.gov.uk/id _ 2004 _ summary _ briefing).

Harwood, P.D.L. (1976) 'Quality of life: ascriptive and testimonial conceptualisations', *Social Indicators Research* 7: 463–76.

Hayes, M.G. (1986) 'Urban decline and deprivation: Liverpool's Relative Position', Technical Study 1, Liverpool: Liverpool City Council.

Held, D. (1991) 'Democracy, the nation-state and the global system', *Economy and Society* 20: 138–72.

Hemphill, L., Berry, J. and McGreal, S. (2004a) 'An indicator-based approach to measuring sustainable urban regeneration performance: Part 1, conceptual foundations and methodological framework', *Urban Studies* 41(4): 725–55.

Hemphill, L., Berry, J. and McGreal, S. (2004b) 'An indicator-based approach to measuring sustainable urban regeneration performance: Part 2, empirical evaluation and case-study analysis', *Urban Studies* 41(4): 727–72.

Herman, R., Ardekani, S.A., Govind, S. and Dona, E. (1988) 'The dynamic characterisation of cities', in J.H. Ausubel and R. Herman (eds) *Cities and Their Vital Systems: Infrastructure – Past, Present, and Future*, Washington, DC: National Academy Press.

Herrschel, T. (1995) 'Local policy restructuring: a comparative assessment of policy responses in England and Germany', *Area* 27(3): 228–41.

Hirschfield, A. (1989) 'The study of poverty and deprivation in developed countries: a selective review of data sources, approaches and analytical techniques', The Urban Research and Policy Evaluation Regional Research Laboratory Working Paper 2, Liverpool: Department of Civic Design, University of Liverpool.

HM [Her Majesty's] Government (1986) *Paying for Local Government*, Cm 9714, London: HMSO.

HM [Her Majesty's] Government (1990) *This Common Inheritance*, Cm 1200, London: HMSO.

HM [Her Majesty's] Government (1998) *Statistics: A Matter of Trust*, a consultation document, London: Stationery Office.

HM [Her Majesty's] Government (1999a) *Building Trust in Statistics*, Cm 4412, London: Stationery Office.

HM [Her Majesty's] Government (1999b) *Modernising Government*, Cm 4310, London: Stationery Office.

HM [Her Majesty's] Government (2000) *Local Government Act 2000 Part 1*, London: TSO.

HM [Her Majesty's] Government (2004) *Planning and Compulsory Purchase Act 2004*, London: HMSO.

HM Treasury (2000a) *Framework for National Statistics*, 1st edn, London: Treasury.

HM Treasury (2000b) *Initial Scope of National Statistics*, London: Treasury.

Hodge, I., Dunn, J., Monk, S. and Kiddle, C. (2000) 'An exploration of "bundles" as indicators of rural disadvantage', *Environment and Planning A* 32: 1869–87.

Hoernig, H. and Seasons, M. (2004) 'Monitoring of indicators in local and regional planning practice: concepts and issues', *Planning Practice and Research* 19(1): 81–99.

Holtermann, S. (1975) 'Areas of urban deprivation in Great Britain: an analysis of 1971 Census data', *Social Trends* 6: 33–47.

Horn, R.V. (1993) *Statistical Indicators for the Economic and Social Sciences*, Cambridge: Cambridge University Press.

Houghton, M. (1997) 'Performance indicators in town planning: much ado about nothing?' *Local Government Studies* 23(2): 1–13.

Housing Corporation (1988) *Housing Needs Indicators*, internal document prepared by the Programme and Operational Support Division, June 1988, London: Housing Corporation.

Hughes, J.T. (1991) 'Evaluation of local economic development: a challenge for policy research', *Urban Studies* 28(6): 909–18.

Innes, J.E. (1990) *Knowledge and Public Policy: The Search of Meaningful Indicators*, New Brunswick, NJ: Transaction Publishers.

Innes, J.E. (2002) 'Improving policy making with information', *Planning Theory and Practice* 3(1): 102–4.

Innes, J.E. and Booher, D.E. (2000) 'Indicators for sustainable communities: a strategy for building on complexity theory and distributed intelligence', *Planning Theory and Practice* 1(2): 173–86.

Institute for Research on Poverty (1998) 'Revising the poverty measure', *Focus* 19(2): 1–55.

Jarman, B. (1983) 'Identification of underprivileged areas', *British Medical Journal* 286: 1705–9.

Jarman, B. (1984) 'Underprivileged areas: validation and distribution of score', *British Medical Journal* 289: 1578–92.

JCCI [Jacksonville Community Council Inc.] (2005) JCCI website (www.jcci.org).

Jensen-Butler, C. (1995) 'A theoretical framework for analysis of urban economic policy', paper presented at the Regional Studies Association European Conference: Regional Futures: Past and Present, East and West, Gothenburg, Sweden, 6–9 May.

Jessop, B. (1993) 'Towards a Schumpeterian workfare state? Preliminary remarks on post-Fordist political economy', *Studies in Political Economy* 40: 7–40.

Jessop, B. (1994) 'Post-Fordism and the state', in A. Amin (ed.) *Post-Fordism: A Reader*, Oxford: Blackwell.

Johnson, J.D. and Rasker, R. (1995) 'The role of economic and quality of life values in rural business location', *Journal of Rural Studies* 11(4): 405–16.

Johnston, R., Voas, D. and Poulsen, M. (2003) 'Measuring spatial concentration: the use of threshold profiles', *Environment and Planning B: Planning and Design* 30: 3–14.

Joshi, H. (2001) 'Is there a place for area-based initiatives?' *Environment and Planning A* 33: 1349–52.

Kearns, A. and Forrest, R. (2000) 'Social cohesion and multilevel urban governance', *Urban Studies* 37(5–6): 995–1017.

Kearns, A., Gibb, K. and Mackay, D. (2000) 'Area deprivation in Scotland: a new assessment', *Urban Studies* 37(9): 1535–59.

Kearns, A. and Turok, I. (2003) *Sustainable Communities: Dimensions and Challenges*, report prepared for the Economic and Social Research Council and the Office of the Deputy Prime Minister, Glasgow: Department of Urban Studies, University of Glasgow.

Kleinman, M. (1999) 'There goes the neighbourhood: area policies and social exclusion', *New Economy* 6: 188–92.

Knox, P. (1978) 'Territorial social indicators and area profiles', *Town Planning Review* 49: 75–83.

Knox, P. (1985) 'Disadvantaged households and areas of deprivation: microdata from the 1981 Census of Scotland', *Environment and Planning A* 17: 413–25.

Kresl, P.K. and Singh, B. (1999) 'Competitiveness and the urban economy: twenty-four large US metropolitan areas', *Urban Studies* 36 (5–6): 1017–27.

Land, K.C. (1975) 'Theories, models and indicators of social change', *International Social Science Journal* 27(7): 7–37.

Land, K.C. (2000) 'Social indicators and quality of life', in E.F. Borgatta and R.V. Montgomery (eds) *Encyclopedia of Sociology*, rev. edn, New York: Macmillan.

Land, K.C. and Felson, M. (1976) 'A general framework for building dynamic macro social indicator models: including an analysis of changes in crime rates and police expenditures', *American Journal of Sociology* 82: 565–604.

Land, K.C. and McMillen, M.M. (1980) 'A macrodynamic analysis of changes in mortality indexes in the United States, 1947–75: some preliminary results', *Social Indicators Research* 7: 1–46.

Landis, J.D. and Sawicki, D.S. (1988) 'A planner's guide to the Places Rated Almanac', *American Planning Association Journal* 54(3): 336–46.

Lazarsfeld, P. (1959) 'Problems in methodology', in R.K. Merton (ed.) *Sociology Today*, New York: Basic Books.

Lazarsfeld, P. (1970) *La Philosophie des Sciences Sociales*, Paris: Gallimard.

Lee, P., Murie, A. and Gordon, D. (1995) *Area Measures of Deprivation: A Study of Current Methods and Best Practices in the Identification of Poor Areas in Great Britain*, a report to the Joseph Rowntree Foundation, Birmingham: Centre for Urban and Regional Studies, University of Birmingham.

Lever, W. and Turok, I. (1999) 'Competitive cities: introduction to the review', *Urban Studies* 36(5–6): 791–3.

Levett, R. (1998) 'Sustainability indicators – integrating quality of life and environmental protection', *Journal of Royal Statistical Society A* 161: 291–302.

LGMB (1995) *Sustainability Indicators Research Project: Consultants' Report of the Pilot Phase*, Luton: Local Government Management Board.

Little, A. and Mabey, C. (1972) 'An index for designation of Educational Priority Areas', in A. Shonfield and S. Shaw (eds) *Social Indicators and Social Policy*, published for the Social Science Research Council, London: Heinemann Educational Books.

Liu, B.-C. (1976) *Quality of Life Indicators in US Metropolitan areas: A Statistical Analysis*, New York: Praeger.

Longford, N. (2001) 'Comment on DETR Indices of Deprivation ID2000', London: Royal Statistical Society, web page (http://stats.lse.ac.uk/galbrait/indices/).

Loomba, A.P.S. and Johannessen, T.B. (1997) 'Malcolm Baldrige National Quality Award critical issues and inherent values', *Benchmarking for Quality Management and Technology* 4(1): 59–73.

Lupton, R. and Power, A. (2002) 'Social exclusion and neighbourhoods', in J. Hills, J. Le Grand and D. Piachaud (eds) *Understanding Social Exclusion*, Oxford: Oxford University Press.

Lupton, R. and Power, A. (2004) 'What we know about neighbourhood change: a summary', paper prepared for ESRC/ODPM 'Understanding Neighbourhood Change' seminar, at Central Hall, London, 30 March.

McCulloch, A. (2001) 'Ward-level deprivation and individual social and economic outcomes in the British Household Panel Study', *Environment and Planning A* 33: 67–84.

Maclaren, V.W. (1996) 'Urban sustainability reporting', *American Planning Association Journal*, spring 1996: 184–201.

Macnaghten, P., Grove-White, R., Jacobs, M. and Wynne, B. (1995) *Public Perceptions and Sustainability in Lancashire: Indicators, Institutions, Participation*, Preston: Lancashire County Council.

MacRae, D. (1985) *Policy Indicators: Links between Social Science and Public Debate*, Chapel Hill: University of North Carolina Press.

Mair, A. (1993) 'New growth poles? Just-in-time manufacturing and local economic development strategy', *Regional Studies* 27(3): 207–21.

Malecki, E.J. and Bradbury, S.L. (1992) 'R&D facilities and professional labour: labour force dynamics in high technology', *Regional Studies* 26(2): 123–36.

Mannheim, K. (1936) *Ideology and Utopia: An Introduction to the Sociology of Knowledge*, trans. L. Wirth and E.A. Shils, New York: Harcourt, Brace & World.

Markusen, A., Hall, P. and Glasmeier, A. (1986) *High Tech America*, Boston: Allen & Unwin.

Martin, C.J. (1989) 'Researching the obvious and influencing the influentials', *Local Government Policy Making* 16(3): 47–52.

Martin, D., Senior, M.L. and Williams, H.C.W.L. (1994) 'On measures of deprivation and the spatial allocation of resources for primary healthcare', *Environment and Planning A* 26: 1911–29.

Massam, B.H. (1993) *The Right Place: Shared Responsibility and the Location of Public Facilities*, Harlow: Longman.

Miles, I. (1985) *Social Indicators for Human Development*, London: Frances Pinter.

Morris, H. (2003) 'MP turns against South East homes', *Planning*, 2 May 2003: 1.

Morris, R. and Carstairs, V. (1991) 'Which deprivation? A comparison of selected deprivation indexes', *Journal of Public Health Medicine* 13(4): 318–26.

Morrison, N. and Pearce, B. (2000) 'Developing strategic indicators for evaluating the effectiveness of the UK land use planning system', *Town Planning Review* 71(2): 191–212.

Myers, D. (1988) 'Building knowledge about quality of life for urban planning', *Journal of the American Planning Association* 54(3): 347–58.

National Audit Office (1990) *Regenerating the Inner Cities*, HC 169, London: HMSO.

National Health Service in Scotland [NHS Scotland] (2001) 'The legal test of Regulation 5 (10)', website (www.show.scot.nhs.uk/shsc/nationalappealpanel/thelegaltest rmacdonald091002.doc).

NEI [Netherlands Economic Institute] (1992) *New Location Factors for Mobile Investment in Europe*, Brussels: DGXVI, Commission of the European Communities.

Newcastle upon Tyne City Council (1986) *Audit Commission – Authority Profile: Use of Z-scores Indicators*, report by Head of Policy Services to Performance, Review and Efficiency Sub-Committee, 4 April 1986, Newcastle upon Tyne: Newcastle upon Tyne City Council.

Noble, M., Firth, D., Dibben, C., Lloyd, M., Smith, G. and Wright, G. (2001) *Deprivation in London: An Alternative to ID2000*, unpublished report to the Greater London Authority, City Hall, The Queen's Walk, London SE1 2AA.

Noble, M., Penhale, B., Smith, G. and Wright, G. (1999) *Measuring Multiple Deprivation at the Local Level*, Index of Deprivation 1999 Review, Oxford: University of Oxford, Department of Applied Social Studies and Social Research.

Noble, M., Penhale, B., Smith, G., Wright, G., Firth, D. and Payne, C. (2000a) *Report for Formal Consultation Stage 2: Methodology for an Index of Multiple Deprivation*, Index of Deprivation 1999 Review, Oxford: University of Oxford, Department of Applied Social Studies and Social Research.

Noble, M., Smith, G., Penhale, B., Wright, G., Dibben, C., Owen, T. and Lloyd, M. (2000b) *Measuring Multiple Deprivation at the Small Area Level: The Indices of Deprivation 2000*, London: Department of the Environment, Transport and the Regions, Regeneration Series.

Noble, M., Wright, G., Dibben, C., Smith, G., McLennan, D., Anttila, C., Barnes, H., Mokhtar, C., Noble, S., Avenell, D., Gardner, J., Covizzi, I. and Lloyd, M. (2004) *The English Indices of Deprivation 2004 (Revised)*, London: Office of the Deputy Prime Minister.

ODPM [Office of the Deputy Prime Minister] (2002a) *Monitoring Regional Planning Guidance: Good Practice Guidance on Targets and Indicators*, London: Office of the Deputy Prime Minister.

ODPM [Office of the Deputy Prime Minister] (2002b) 'The development of town and city indicators database', *Urban Research Summary* 3, London: Office of the Deputy Prime Minister.

ODPM [Office of the Deputy Prime Minister] (2003a) *Sustainable Communities: Building for the Future*, London: Office of the Deputy Prime Minister.

ODPM [Office of the Deputy Prime Minister] (2003b) *Consultation Paper on Draft Planning Policy Statement 11 (PPS11) – Regional Planning*, London: Office of the Deputy Prime Minister.

ODPM [Office of the Deputy Prime Minister] (2004a) *Making It Happen: The Northern Way*, London: Office of the Deputy Prime Minister.

ODPM [Office of the Deputy Prime Minister] (2004b) *Consultation Paper on Planning Policy Statement 1: Creating Sustainable Communities*, London: Office of the Deputy Prime Minister.

ODPM [Office of the Deputy Prime Minister] (2004c) *Planning Policy Statement 11: Regional Spatial Strategies*, London: Office of the Deputy Prime Minister.

ODPM [Office of the Deputy Prime Minister] (2004d) *Planning Policy Statement 12: Local Development Frameworks,* London: Office of the Deputy Prime Minister.

ODPM [Office of the Deputy Prime Minister] (2004e) 'Developing a Town and City Indicators Database', *Urban Research Summary*, November 17, London: Office of the Deputy Prime Minister.

ODPM [Office of the Deputy Prime Minister] (2005a) *Core Output Indicators for Regional Planning*, London: Office of the Deputy Prime Minister.

ODPM [Office of the Deputy Prime Minister] (2005b) *Local Development Framework Monitoring: A Good Practice Guide*, London: Office of the Deputy Prime Minister.

ODPM [Office of the Deputy Prime Minister] (2005c) *Planning Policy Statement 1: Delivering Sustainable Development*, London: Office of the Deputy Prime Minister.

Olson, M., Jnr (1969) 'Rapport social, indicateurs sociaux, comptes sociaux', *Analyse et prévision*, February 1969.

ONS [Office for National Statistics] (1998) *Regional Trends*, 33, London: Stationery Office.

ONS [Office for National Statistics] (2002) *Access Small Areas: The Newsletter for Neighbourhood Statistics*, November, London: ONS.

Openshaw, S. (1984) *The Modifiable Areal Unit Problem*, Norwich: GeoBooks.

Openshaw, S., Sforzi, F. and Wymer, C. (1985) 'A multivariate classification of individual household census data for Italy', *Papers of the Regional Science Association* 58: 113–25.

Pacione, M. (1995) 'The geography of deprivation in rural Scotland', *Transactions* 20(2): 173–92.

PA Cambridge Economic Consultants (1990) *Indicators of Comparative Regional/Local Economic Performance and Prospects*, London: HMSO.

PAG [Planning Advisory Group] (1965) *The Future of Development Plans*, Planning Advisory Group, London: HMSO.

Pattie, C. (2001) 'On reinvented wheels', *Environment and Planning A* 33: 1353–6.

Percy-Smith, J., Sanderson, I., Dowson, L. and Hutchinson, J. (2000) *Enhancing Research Capacity: the Importance of Research in Modern Local Government*, London: the Local Government Association.

Peck, J. and Tickell, A. (1995) 'Jungle law breaks out: neoliberalism and global-local disorder', *Area* 26(4): 317–32.

Pieda (1995) *Local Economic Audits: A Practical Guide*, Sheffield: Employment Department.

Pinder, J. (1980) 'Policy and research II', *Policy Studies* 1: 8–13.

PIU [Performance and Innovation Unit] (2000a) *Adding It Up*, London: Performance and Innovation Unit.

PIU [Performance and Innovation Unit] (2000b) *Reaching out: The Role of Central Government at Regional and Local Level*, London: Performance and Innovation Unit.

Porter, M.E. (1990) *The Competitive Advantage of Nations*, London: Macmillan.

Post, W. and Wieringa, K. (1997) *DAFIA-1: Data-Flow Analysis for Integrated Assessments*, Brussels: Joint Research Centre of the European Environment Agency.

Premus, R. (1982) *Location of High Technology Firms and Regional Development*, Washington, DC: Government Printing Office.

Ragin, C. (1987) *The Comparative Methods: Moving beyond Qualitative and Quantitative Strategies*, Berkeley: University of California Press.

Rhind, D. (1983) 'Mapping census data', in D. Rhind (ed.) *A Census User's Handbook*, London: Methuen.

Rich, R.F. (1981) *Social Science Information and Public Policy Making*, San Francisco: Jossey-Bass.

Rittel, H. and Webber, M. (1973) 'Dilemmas in a general theory of planning', *Policy Sciences* 4: 155–69.

Roback, J. (1982) 'Wages, rents and the quality of life', *Journal of Political Economy* 90: 1257–8.

Roberts, P., Collis, C. and Noon, D. (1990) 'Local economic development in England and Wales: successful adaptation of old industrial areas in Sedgefield, Nottingham and Swansea', in W.B. Stöhr (ed.) *Global Challenge and Local Response*, London: Mansell Publishing.

Robson, B. (2003) 'Evaluation approaches', a presentation to the Northwest Development Agency's Evidence Based Policy Development for Regeneration Schemes Masters Classes, 22 January 2003.

Robson, B., Barr, R., Wong, C., Bowers, K., Bradford, M. and Mazzi, M. (2003) *Updating and Revising the Welsh Index of Deprivation*, Study Report, Cardiff: Local Government Data Unit of Wales.

Robson, B., Bradford, M. and Tomlinson, R. (1998) *Updating and Revising the Index of Local Deprivation*, London: Department of the Environment.

Robson, B., Bradford, M. and Tye, R. (1995) *The 1991 Deprivation Index: A Matrix of Results*, London: HMSO.

Robson, B., Deas, I., Mazzei, M., Whyte, D., Hirschfield, A., Bowers, K. and Johnson, S. (2002) *Developing New Approaches to the Measurement of Deprivation*, London: Association of London Government.

Robson, B., Parkinson, M., Boddy, M. and Maclennan, D. (2000) *The State of English Cities*, London: DETR.

Rogerson, R. (1999) 'Quality of life and city competitiveness', *Urban Studies* 36(5–6): 969–85.

Rogerson, R., Findlay, A. and Morris, A. (1987) 'The geography of quality of life', Department of Geography Occasional Paper Series No. 22, Glasgow: University of Glasgow.

Rogerson, R., Findlay, A., Morris, A. and Coombes, M. (1989) 'Indicators of quality of life: some methodological issues', *Environment and Planning A* 21: 1655–66.

Room, G. (ed.) (1995) *Beyond the Threshold: The Measurement and Analysis of Social Exclusion*, Bristol: Policy Press.

Rose, R. (1972) 'The market for policy indicators', in A. Shonfield and S. Shaw (eds) *Social Indicators and Social Policy*, published for the Social Science Research Council, London: Heinemann Educational Books.

Rosen, S. (1979) 'Wage-based indexes of urban quality of life', in P. Mieszkowski and M. Straszheim (eds) *Current Issues in Urban Economics*, Baltimore, MD: Johns Hopkins University Press.

RSS [Royal Statistical Society] (1995) *Report of the Working Party on the Measurement of Unemployment in the UK*, London: Royal Statistical Society.

RSS [Royal Statistical Society] (1998) *The Response of the Royal Statistical Society to the Government's Consultation Document 'Statistics: A Matter of Trust'*, London: Royal Statistical Society.

RSS [Royal Statistical Society] (1999a) 'Building trust on shaky foundations: government's statistical proposals are deeply flawed', 18 October 1999, The Royal Statistical Society web page (www.rss.org.uk/archive/whiteresp.html).

RSS [Royal Statistical Society] (1999b) 'RSS response to the government's White Paper: Building Trust in Statistics', 3 December 1999, The Royal Statistical Society web page (www.rss. org.uk/archive/whiteresp.html).

Sackman, H. (1974) *Delphi Assessment: Expert Opinion, Forecasting and Group Procedure*, Santa Monica, CA: Rand Corporation.

Sammons, P., Thomas, S., Mortimore, P., Owen, C. and Pennell, H. (1994) *Assessing School Effectiveness: Developing Measures to Put School Performance in Context*, London: Institute of Education.

Sampson, R.J. (2004) 'How does community context matter? Social mechanisms and the explanation of crime', paper presented at the 2004 ESRC Cambridge Social Context of Pathways in Crime Conference, website (www.scopic.ac.uk/communitycontext.htm).

Sanderson, I., Percy-Smith, J. and Dowson, L. (2001) 'The role of research in modern local government', *Local Government Studies* 27(3): 59–78.

Saunders, J. (1998) 'Weighted census-based deprivation indices: their use in small areas', *Journal of Public Health Medicine* 20(3): 253–60.

Sawicki, D.S. (2002) 'Improving community indicator systems: injecting more social science into the folk movement', *Planning Theory and Practice* 3(1): 13–32.

Sawicki, D. and Flynn, P. (1996) 'Neighbourhood indicators: a review of the literature and an assessment of conceptual and methodological issues', *Journal of the American Planning Association* 62(2): 165–83.

Sayer, A. (1992) 'Radical geography and marxist political economy: towards a re-evaluation', *Progress in Human Geography* 16: 343–60.

Schaller, N. (1993) 'The concept of agricultural sustainability', *Agriculture, Ecosystems and Environment* 46: 89–97.

Schmenner, R. (1982) *Making Business Location Decisions*, Englewood Cliffs, NJ: Prentice-Hall.

Schuessler, K.F. and Fisher, G.A. (1985) 'Quality of life research and sociology', *Annual Review of Sociology* 11: 129–49.

SDRC [Social Disadvantage Research Centre] (2004) 'Understanding neighbour-hood change: challenges and potential for using and combining data at neighbourhood level', paper prepared by SDRC, University of Oxford for the ESRC/ODPM 'Understanding Neighbourhood Change' seminar, at Central Hall, London, 30 March.

Seebohm Committee (1968) *Report of the Committee on Local Authority and Allied Personal Social Services*, Cmnd 3703, London: HMSO.

SEU [Social Exclusion Unit] (2000) *Better Information*, Social Exclusion Unit Policy Action Team Report 18, London: Social Exclusion Unit.

Shah, R. (2004) 'Commission on Sustainable Development Indicators of Sustainable Development – recent developments and activities', paper presented at the Assess-ment of Sustainability Indicators Workshop, 10–14 May 2004, Prague, Czech Republic.

Sheldon, E.B. and Moore, W.E. (eds) (1968) *Indicators of Social Change: Concepts and Measurements*, New York: Russell Sage Foundation.

Simpson, S. (1996) 'Resource allocation by measures of relative social need in geographical areas: the relevance of the signed chi-square, the percentage, and the raw count', *Environment and Planning A* 28: 537–54.

Skelcher, C. (1982) 'Corporate planning in local government', in S. Leach and J.D. Stewart (eds) *Approaches in Public Policy*, London: Allen & Unwin.

Skoro, C.L. (1988) 'Rankings of state business climates: an evaluation of their useful-ness in forecasting', *Economic Development Quarterly* 2(2): 138–52.

Smith, D. (1973) *Geography of Social Well-being*, New York: McGraw-Hill.

Smith, T.W. (1980/1) 'Social indicators: a review essay', *Journal of Social History* 14: 739–47.

Snape, D. and Boddy, M. (1996) 'Local government research and the policy process: a threatened relationship?' *Local Government Policy Making* 22(5): 39–43.

SNS [Scottish Neighbourhood Statistics] (2004) 'Scottish Index of Deprivation', Scottish Neighbourhood Statistics website (www.scotland.gov.uk/stats/neighbours/ tables/index.asp).

Stewart, R. (1995a) 'Indicating the possible', *Planning Week* 6(9): 12–13.

Stewart, R. (1995b) *National Review of the Role of Research in Local Government*, Report of the Advisory Group, Wokingham: Local Authority Research and Intelligence Association.

Stewart, M. (2000) 'Reflections on neighbourhood policy', the King's Fund Regeneration and Mental Health, Briefing 3, May 2000: 3.

Stöhr, W.B. (1987) 'Regional economic development and the world economic crisis', *International Social Science Journal* 112: 187–97.

Stone, R. (1971) *Demographic Accounting and Model-Building*, Paris: Organisation for Economic Cooperation and Development.

Stover, M.E. and Leven, C.L. (1992) 'Methodological issues in the determination of the quality of life in urban areas', *Urban Studies* 29(5): 737–54.

Sustainable Seattle (2005) 'Indicators of Sustainable Community' website (www.sustainableseattle.org).

Swain, D. and Hollar, D. (2003) 'Measuring progress: community indicators and their quality of life', *International Journal of Public Administration* 26(7): 789–814.

Taylor, C.L. (1981) 'Progress towards indicator systems: an overview', in C.L. Taylor (ed.) *Indicator Systems for Political, Economic and Social Analysis*, Cambridge, MA: Gunn & Hain.

Taylor, J. (1993) 'An analysis of the factors determining the geographical distribution of Japanese manufacturing investment in the UK 1984–91', *Urban Studies* 30(7): 1209–24.

Taylor, M. (2000) 'Communities in the lead: power, organisational capacity and social capital', *Urban Studies* 37(5): 1019–35.

Theotherpages (2004) Quotations webpage (www.theotherpages.org/topic-dl.html).

The Concise Reference Encyclopedia and Dictionary (1987) Compiled in association with Oxford University Press, London: Bay Books.

Thomas, M.J. (1982) 'The procedural planning theory of A. Faludi', in C. Paris (ed.) *Critical Readings in Planning Theory*, Oxford: Pergamon Press.

Tomlinson, M. and Kelly, G. (2003) 'What's the use of the Noble index? Theories, methods and applications', in E. McLaughlin and G. Kelly (eds) *Anti-Poverty Strategies in Ireland, North and South*, Belfast: Department for Social Development and Social Security Research, Queen's University of Belfast.

Townsend, P. (1987) 'Deprivation', *Journal of Social Policy* 16(2): 125–46.

Turok, I. and Edge, N. (1999) *The Job Gap in Britain's Cities: Employment Loss and Labour Market Consequences*, Bristol: Policy Press.

UN Statistical Office (1975) *Towards a System of Social and Demographic Statistics*, Studies in Methods, Series F, No. 18, New York: UN Statistical Office.

UNCED [United Nations Commission on Environment and Development] (1992) *Agenda 21*, Conches, Switzerland: UNCED.

UNCSD [United Nations Commission on Sustainable Development] (1996) *Indicators of Sustainable Development: Framework and Methodologies*, New York: UNCSD.

UNCSD [United Nations Commission on Sustainable Development] (2001) *Indicators of Sustainable Development: Guidelines and Methodologies*, New York: UNCSD.

US Department of Health, Education and Welfare (1969) *Toward a Social Report*, Washington, DC: US Government Printing Office.

van Gestel, T. and Faludi, A. (2005) 'Towards a European Territorial Cohesion Assessment Network: a bright future for ESPON?' *Town Planning Review* 76(1): 81–92.

Ward, M. (1980) 'Progress towards indicator systems: an overview', in C.L. Taylor (ed.) *Indicator Systems for Political, Economic and Social Analysis*, Cambridge, MA: Gunn & Hain.

Weber, M. (1964) *Basic Concepts in Sociology*, New York: The Citadel Press.

Webber, R.J. (1989) 'Using multiple data sources to build an area classification system: operational problems encountered by MOSAIC', *Journal of the Market Research Society* 31: 103–9.

Webber, R. and Craig, J. (1976) 'Which local authorities are alike?' *Population Trends* 5: 13–19.

Weiss, C.H. (1979) 'The many meaning of research utilization', *Public Administration Review*, September/October: 426–31.

Weiss, C.H. (1995) 'The haphazard connection: social science and public policy', *International Journal of Education Research* 23(2): 137–50.

Weiss, C.H. (with M. Bucuvalas) (1980) *Social Science Research and Decision-Making*, New York: Columbia University Press.

Westfall, M.S. and de Villa, V.A. (2001) *Urban Indicators for Managing Cities: Cities Data Book*, Manila: Asian Development Bank.

WMEB (1993) *Study of Existing Evidence on Local Economic Development and TECs*, a final report to the Department of Employment, Birmingham: WMEB Consultants.

Williams, S. (1980) 'Planning and research: I', *Policy Studies* 1: 2–7.

Wong, C. (1993) 'Towards better practice in planning information collection', *Town and Country Planning* 62(10): 279–81.

Wong, C. (1995) 'Developing quantitative indicators', in R. Hambleton and H. Thomas (eds) *Urban Policy Evaluation*, London: Paul Chapman.

Wong, C. (1998a) 'Determining factors for local economic development: the perception of practitioners in the North West and Eastern Regions of the UK', *Regional Studies* 32: 707–20.

Wong, C. (1998b) 'Inter-relationships between key actors in local economic development', *Environment and Planning C: Government and Policy* 16: 463–81.

Wong, C. (2000) 'Indicators in use: challenges to urban and environmental planning in Britain', *Town Planning Review* 71(2): 213–39.

Wong, C. (2001) 'The relationship between quality of life and local economic development: an empirical study of local authority areas in England', *Cities* 18: 25–32.

Wong, C. (2002a) 'Developing indicators to inform local economic development in England', *Urban Studies* 39(10): 1833–63.

Wong, C. (2002b) 'The crossroads of community indicators', *Planning Theory and Practice* 3(2): 259–60.

Wong, C. (2002c) *Development of Town and City Indicators: Review, Interpretation and Analysis*, a final report to the Department for Transport, Local Government and the Regions, Liverpool: Policy Evaluation and Analysis Research Laboratory, University of Liverpool.

Wong, C. (2003) 'Indicators at the crossroads: ideas, methods and applications', *Town Planning Review* 74(3): 253–79.

Wong, C., Baker, M. and Kidd, S. (2005) *Local Development Framework Monitoring: A Good Practice Guide*, London: Office of the Deputy Prime Minister.

Wong, C., Jeffery, P., Green, A., Owen, D., Coombes, M. and Raybould, S. (2004) *Town and City Indicators Database Project*, London: Office of the Deputy Prime Minister.

Wong, C., Ravetz, J. and Turner, J. (2000) *The United Kingdom Spatial Planning Framework*, London: The Royal Town Planning Institute.

World Bank (1995) *Social Indicators of Development*, Washington, DC: World Bank.

World Bank (2003) *World Development Indicators 2003*, Washington, DC: World Bank.

World Commission on Environment and Development (1987) *Our Common Future*, Oxford: Oxford University Press.

Worral, L. (1991) 'Local and regional information systems for public policy', in M.J. Healey (ed.) *Economic Activity and Land Use: The Changing Information Base for Local and Regional Studies*, London: Longman.

WSA (1999) *Review of the Local Government Research Programme*, a final report to the Department of the Environment, Transport and the Regions, London: William Solesbury and Associates Research Management Consultancy.

Zapf, W. (1981) 'Applied social reporting: a social indicators system for West German society', in C.Y. Taylor (ed.) *Indicator Systems for Political, Economic and Social Analysis*, Cambridge, MA: Gunn & Hain.

INDEX

ABI *see* Annual Business Inquiry
ACORN (*A Classification of Residential Neighbourhoods*) 71–2
Adding It Up: report (PIU 2000a) 37; website 38
Agenda 21 164–6, 187, 190
Allsopp Report 61–4, 68–69, 159–61, 185–6, 187
analysis: analytical indicator bundle approach 96–9; area-level 127–9; cluster 72, 87; data 95–103; factor 86; framework *see* analytical structuring; multi-criteria 87; regression 86; trend 99–100, 101–2
analytical principles 111–12
analytical structuring 106, 109–12
Annual Business Inquiry (ABI) 68–69
ARO *see* Association of Regional Observatories
Asian Development Bank 142
Association of Regional Observatories (ARO) 60–2

basic research – policy application nexus 17
benchmarking 98–100
Best Value 36–8, 45, 46, 70, 75, 79
Better Information report (SEU 2000) 48, 55, 56, 67, 78, 140, 185
boundaries, administrative 51, 93, 130
Breadline Britain Index 123, 129
Building Partnerships for Prosperity White Paper (DETR 1997) 58, 62, 150–1, 161
Building Trust in Statistics White Paper (HM Government 1999a) 49, 52–3, 56, 64

Capital Challenge (1990s) 32
centralisation and decentralisation 40
chi-square values 116, 123, 125, 131
City Challenge (1991) 32

Classification of Residential Neighbourhoods, A *see* ACORN
cluster analysis *see* analysis: cluster
community indicators movement, the 2, 5, 171–2, 188–9
conceptual consolidation 105–8
Countryside Agency 126, 136; *see also* Rural Development Commission

data 68–80; analysis and interpretation 95–102; compilation according to administrative boundaries 74–6; crime 72; employment 68–69; measurement difficulties 78–9; spatial 74–7; survey-based 72–4
data agencies 26
Data Protection Act (DPA) 70–1, 140
decentralisation and centralisation 40
definitions: deprivation 106, 121–4; disadvantage 124, 125; indicators 3–4; poverty 122; quality of life 143–4; social exclusion 124; sustainable development 146–7, 164
deprivation: key indices since 1980s 123; multiple 124–5, 128–9
Deprivation Index (DoE 1981) *see* Index of Deprivation (DoE 1981)
deprivation indicators 46, 55, 74, 122–41, 186
deprivation indices: area-based measures and policies 127–9, 186; criticisms 140–1; measurement issues 133–4; political meanings 134–9; review and development 135–9; spatial scales and administrative areas 130–1; urban-rural dimension 126–7; vulnerable groups 129–30; weightings, statistical standardisation and transformation 131–3
devolution of policy 160–1
disadvantage *see* definitions, disadvantage